The Historical Ecology of
Malaria in Ethiopia

Ohio University Press Series in Ecology and History
James L. A. Webb, Jr., Series Editor

The Historical Ecology of Malaria in Ethiopia

Deposing the Spirits

James C. McCann

OHIO UNIVERSITY PRESS
ATHENS

Ohio University Press, Athens, Ohio 45701
ohioswallow.com
© 2014 by Ohio University Press
All rights reserved

Printed in the United States of America
Ohio University Press books are printed on acid-free paper. ♾ ™

20 19 18 17 16 15 14 5 4 3 2 1

Library of Congress Cataloging-in-Publication Data

McCann, James, 1950– author.
The historical ecology of malaria in Ethiopia : deposing the spirits / James C. McCann.
   pages cm. — (Series in ecology and history)
ISBN 978-0-8214-2146-8 (hc : alk. paper) — ISBN 978-0-8214-2147-5 (pb : alk. paper) —
ISBN 978-0-8214-4513-6 (pdf)
1. Malaria—Ethiopia—History. 2. Malaria—Ethiopia—Prevention. I. Title. II. Series: Ohio
University Press series in ecology and history.
RA644.M2M33 2015
614.5'320963—dc23
                                                                          2015004831

To
the memory of Andrew Spielman

Contents

Illustrations

Figures

Tables

Acknowledgments

Books are long roads, a kind of pilgrimage. Yet one does not journey alone. There are many important fellow travelers along that road, near and afar, who make the moving along a valuable venture. This road began in 2003 with an unexpectedly productive collaboration with the late Dr. Andrew Spielman from the Harvard School of Public Health. Andy was a great and inquisitive malariologist who explored the many contexts of malaria and inspired many in that field. Researchers in his lab and their careers are part of his enduring legacy. Those colleagues have played a major role in this book by teaching me about malaria, entomology, and the ecology of this disease. That group includes: Tony Kiszewski, Richard Pollack, Rebecca Robich, and Yemane Ye-Ebiyo. Medical entomologist Rich Pollack replaced Andy, after his death, as my coinvestigator on the Rockefeller grant that supported our work for five years. These colleagues were my teachers on malaria, and I enjoyed watching them learn about history, agriculture, and the social context of the disease as we worked together in the field. We argued, laughed, and learned together with our Ethiopian colleagues. Rich Pollack has also been a careful and critical reader of this manuscript.

In Ethiopia, our team benefited fundamentally from the insights and hard work of my friend Asnakaw Kebede, a brilliantly effective malaria specialist who shared his years of experience and sense of balance in our fieldwork together. His skills as a local diplomat and friend match his talents as a researcher. Yihenew Tesfaye Alemu joined us to learn, organize, and share thoughts as the project gathered evidence and broadened its perspective. I would guess that each of these gifted observers has come to a belief that malaria is as much ecology as it is bioscience.

Colleagues at the African Studies Center at Boston University were helpful regarding field research and in sharing ideas. These included Michael DiBlasi and Magaly Koch, who brought insights from archaeology, remote sensing, and mapping. Joanne Hart's spirit kept the books in order, as she has on so many projects. Students also joined the group for recording interviews (Sofia Abba Jebel) and collecting visual images (Molly Williams). Caroline Smartt began in the back of a classroom for a course on environmental history, eventually brought herself to the field site in Ethiopia, and then quickly proceeded to blend her knowledge of science, humanities, and overall brilliance into the project—an extraordinary young woman. Her dedicated work on gathering evidence from related malaria work was a labor of true devotion that took her to Phi Beta Kappa and thence to the London School of Hygiene and Tropical Medicine on Keppel Street in Bloomsbury, London.

Ethiopia, over five years of project work, was a site for sustaining friendships, long days in muddy fields, extraordinary working meals at the modest Abebe Hotel in Asandabo, and the simple hospitalities of local health stations. Over that wonderful local cuisine there were conversations about illness, health, and the meanings of it all. Key connections at Addis Ababa University also came from Dr. Maresha Fetene and Yeselamwork Sherif, Delnasaw Yewhalaw (in Jimma University), and old friend Semahagn Abate (in Burie). Helmut Kloos was with us in the early days at the Addis Ababa University Guest House and added his special brand of encouragement for me to follow malaria's many paths. Among the colleagues at the Ethiopian Institute for Agricultural Research in Jimma and Addis I could always count on Habte Jifar, Leta Tula, and Tesfa Abate, who organized farmers, supervised field management, and helped set up farmers to receive seed, fertilizer, and cooperation. Ashenafi Kebede in the Omo Nada Ministry of Health smoothed the fieldwork and health data collection at that site. Dennis Friesen at CIMMYT in Addis helped with logistics on the original maize/malaria project. Brown paper pollen bags were more important than one might think. Izabela Orlowska was always a sensitive soul about reading deeper meanings of history and things of life. She knows that. She was the Mozart to my Salieri.

At the World Health Organization library/archive in Geneva, Marie Villemin Partow was a superb finder of things archival on my three visits there. On the French side of the Swiss border, my wonderful hosts were Angela Raven-Roberts, Sue Lautze, and Kuffase Boane, who looked after me in blizzards with champagne chilled in the Divonne snowbank and Beaujolais Brouilly on warmer days. Ambachew Medhin of WHO offered malaria wisdom in Geneva and in Addis Ababa. I also found friends in the family of Dr. Tadesse at the Z Guest House, a place to write and think on the balcony under the bottlebrush tree.

In May and June 2012, I was fortunate to be named the Hiob Ludolf Visiting Professor at the Centre for Ethiopian Studies at the University of

Hamburg. Their superb library collection and scholars of language and old documents offered wonderful insights into the language and contexts of malaria and Ethiopia. The scholars there helped with finding malaria references in Ge'ez texts collected only recently from the field and in finding textual references in their unique collection. I am particularly grateful to Evgenia Sokolinskaia, Magdalena Krzyzanowska, Alessandro Bausi, Angela Müller, and Dirk Bustorf.

At the Pardee Center at Boston University, where I wrote much of this book and where I served as Director ad interimad interim for two years, there was also much inspiration. I am terribly grateful for that superb staff and the extraordinary stimulation of their Longer Range Future perspective. There, Cynthia Barakatt, Theresa White, and Susan Zalkind looked after the place when I was away in the field until I came back to write in the quiet upper garret overlooking the Charles River.

Funding for this project came from a Fulbright-Hays Fellowship, a Boston University start-up grant, and five years of fieldwork support from the Rockefeller Foundation. At Rockefeller, Gary Toennessien was the key animator of the original grant and of the Rockefeller Legacy grant in the final year. David Gillerman of BU's Office of Development and the wonderful Dolores Markey were key to the conception and management of the grants. In the final stage of writing and field visits, I had the benefit of a grant from the John Simon Guggenheim Foundation and Fulbright. Dr. Yohannes Birhanu and the staff at the Public Affairs Office at the U.S. Embassy were always helpful with logistics.

Ohio University Press continues its outstanding work in African studies and its Ecology and History series. James Webb has led that series and has come to be an engaging colleague and friend—and a productive scholar in his own right. Sarah Chava and Deborah Wiseman were careful and effective editors of my somewhat messy draft of the manuscript.

The inevitable errors in judgment, fact, and balance are, of course, mine alone.

Finally, I owe a great debt to my family, including two daughters, who have launched themselves into their own lives, and my partner, Sandi, who have sustained things in so many ways.

Boston and Arlington, Massachusetts, 2014

Malaria's Metaphor

A Chess Game or a Square Dance

> At first the disease attacks the head/mind: The person complains that his head hurts. It then spreads to the shoulders and then to the whole body. . . . First one person is struck down, then another. Five, six people in a house even. Not to mention the neighbors. Everyone knows it is the nidad [malaria, lit. "to catch fire"].
>
> —Asres, healer/magician (Ethiopia, 1953)

> In each house we were able to find three or four patients who complained of subjective symptoms, such as chilling, severe headaches, sweating, pain in the back . . . high fever . . . muddling delirium with coma, ending in death. . . . Since they are so far away from even the simplest clinic, which means no [way] of saving their lives, they are dying like bees in a smoked hive.
>
> —Mogues A., student, Gondar Public Health College (Ethiopia, 1958)

MALARIA IS AN INFECTIOUS DISEASE like no other. It is a dynamic, shape-shifting force of nature that constitutes Africa's most deadly and debilitating vector-borne disease. During its historical coevolution with humankind, malaria has evaded biomedicine's struggles to eradicate it or control its movement. It has mocked efforts by humans to pursue it through single-stranded tactics: applications of DDT, vaccines, chloroquine tablets, and molecular-level genetic manipulations. Despite biomedicine's efforts to find solutions in one-dimensional panacea, malaria survives as a unique human affliction of ecology that justifies a study of its history and its future that accepts its complexity and its local dynamism as one of its fundamental features. Though its impact has a global scale, all malaria is complex, resilient—and local.

Biomedicine's failures to cope with or "eradicate" malaria to date have, in fact, begged a return to a more comprehensive ecological understanding of malaria and its transmission. To paraphrase a political metaphor: All malaria is local, and it is a complex tapestry of nature's forces. Moving within this complexity is a disease organism (plasmodium), a vector (an anopheles mosquito), and a host (the human malaria sufferer).

This book's geographic focus is Ethiopia and that country's kaleidoscope of ecological landscapes. In these places of malaria infection there are fits and starts of policy, human tragedy, and a few triumphs of overcoming periodic crises. Ethiopia's history showcases episodes of malaria and even may point the way forward for the larger global battle.

The goal of this book is to tell an engaging story of human disease ecology that resets our understanding of this deadly disease in human, narrative terms. Medical science is a necessary but not sufficient lens from which to understand the disease. *The Historical Ecology of Malaria in Ethiopia* aims to display the human ecology of the disease with an appreciation of the science of landscape change and the dynamics of a vector-borne infectious disease that has been an enduring element of human history. Malaria continues to shape the tropical and subtropical world in terms of its economic burden and human costs—spiritual as well as physical. Malaria persists and will continue to adjust to changes in both the climate and the human condition. Therein lies a story that is intensely human, but also best explained by ecological science. This book is the final research and writing stage of five years of field study, archival research, and laboratory analysis of mosquitoes, parasites, and human agency in the unstable character of malaria affliction. Ethiopia will be the fascinating stage on which the drama unfolds.

Ethiopia is a landscape that reveals malaria's local history, its global implications, and, perhaps, its future. The region's diverse microecologies defined by elevation and rainfall reflect the astonishing varieties of settings where malaria has coevolved with human settlement and environmental change. In areas along Northeast Africa's borders, where lowland river valleys and warm temperatures nurture mosquito habitat year-round, there is endemic malaria within the local people's bodies, within which constant exposure to the parasites has brought an odd kind of acquired immunity. But those immune populations are small and politically isolated in low borderland regions. In highland areas, where most people live, the altitude and cool temperatures have restricted malaria's effects to "unstable," episodic outbreaks—once or twice a decade, but still especially deadly to adults and to young vulnerable people alike. People's limited exposure has made the "shivering fever" only periodic. This means that the vast majority of people have no acquired immunity to the disease and tend to view it as an affliction of the "other," a disease from outside places and outside forces. In many places Ethiopia's landscapes are in states of transition caused by climate warming, movements of people, and changes in agriculture and land use that invite mosquito populations to shift unexpectedly in species composition or in the abundance of the mosquitoes that are the disease's elusive vectors. Perhaps most importantly, Ethiopia gives us many examples of human landscapes on which malaria plays its deadly game and a lens through which we perhaps can understand its complex strategies as they may appear decades in the future. Malaria may well be a recurrent menace in AD 2060 or for generations thereafter. Or perhaps human medical ingenuity will finally eradicate it globally. But don't bet on it.

To tell the story, we seek a metaphor, or maybe two, to help us imagine a complexity such as malaria's. One such metaphor might be a human-malaria chess game that, over time, becomes a mixing of belief systems and collective behaviors that will shape and describe its causes, prevention, and control. Human science makes a move and malaria responds with a matching move. The end game is control and then eradication. But perhaps a dance metaphor is equally valid, since there is a fluidity of the point and counterpoint between humans and this disease. The human malaria connection is an ever-changing dance, which includes moving forward and back, spinning with partners in tandem or sometimes in opposite directions. Movements of both human beliefs and biological actions. Overlapping meanings. Those belief systems include bioscience and local, preventive behaviors.

One example of changing belief systems in the science world took place in 1880 at a Paris conference when an obscure French colonial army officer named Charles Louis Alphonse Laveran, who had worked in French colonial medical service, presented research about a type of creature—a protozoan—that he believed was the parasite that caused malaria. Laveran had paid his dues as a French army medical officer in malarial colonial Algeria trying to solve the mystery of malaria's cause. After meeting skepticism from science at that first conference, in the next four years Laveran seems to have convinced Louis Pasteur and Italian malariologists about the protozoan cause, and thereby malariology magically merged with germ theory. The earth moved, though it took a quarter century more for bioscience to understand that mosquitoes were the missing link that delivered the infectious agent.[1] But other systems of local knowledge about malaria and its causes existed simultaneously and have continued. Therein lies part of the story.

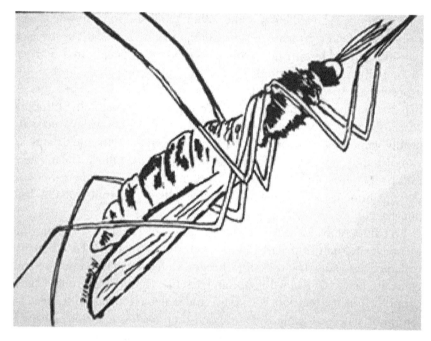

Figure I.1. *An. arabiensis*, Ethiopia's primary vector. (*Source*: Ethiopian Ministry of Health poster. Photo by author.)

Compare that growing European germ theory belief at the turn of the nineteenth century about malaria's cause with that of the Ethiopian healer and freelance cleric named Asres, who understood the ecology of the "shivering fever" (*nidad*) as a product of spirits (*zar*)—perhaps thirty-six or thirty-seven of them, he said, who mischievously afflicted humans at their whim. Asres's belief was closer to that of the dance metaphor, which perhaps was a closer approximation of the swirl of mosquitoes, parasites, and their clever adaptations to humans' steps. But Asres the Ethiopian healer also observed that the fevers occurred in people in some places but not others and to certain types of people but only rarely to others. European travelers quickly learned about malaria from this kind of local knowledge in Ethiopia, which was combined with others' travel experiences in malarial parts of Europe and the Americas. Knowledges merged, swirled, and competed. *Deux a deux* (or "do-si-do," to paraphrase that Cajun dance step).[2]

What kind of mental picture is most helpful in understanding malaria's character as a disease and where a solution for eradication or control might reside? Will malaria's solution derive from the view of complexity that underlies quantum physics (Nobel Prize winner Werner Heisenberg's famous principle of uncertainty) or from gene mapping and genetic manipulation of mosquitoes in laboratories in Geneva or Davis, California? Biosciences punch and mosquitoes/parasites counterpunch.[3] And so it goes.

The approach I take here is to blend stories of place, of time, and of movement. These stories include personal experiences of affliction and of ideas of faith (science, religion, or some combination of beliefs about the natural world). But the stories also include abject failures, local triumphs, the persistence of human will, and nature's own adjustments. Overall, the story is one of collective will in human science, beliefs, and the adaptive genius of coupled human and natural systems. The players in this game of chess—or this dance—vary over time, but by the end of the twentieth century they included mosquitoes, parasites, lab scientists, fieldworkers, sprayers of DDT and other antimosquito products, farmers, and well-meaning international consultants. The fields of play were not so much chessboards as they were laboratories, puddles with cattle hoofprints, pits dug for harvesting clay, houses without window screens, farm field edges, and so on.

Ultimately, the story of malaria, its past, and its future is one in search of a metaphor for a struggle within ecology, an image that evokes movement, adaptation, and complexity. For that the dance offers the appropriate image, no?

What Is Malaria Ecology?

Malaria is a disease that relies on a complicated series of interactions among a parasite-vector-host triad—plasmodium pathogens (there are five kinds that affect people), anopheline mosquito vectors, and bloody meals that the mosquitoes take from both humans and other animals. Changes in the environment/ecology of any of these factors—of their dance—can influence the disease over time, reducing or promoting the frequency of contacts among these players. Who gets malaria, and when? That complexity is both a fact of nature and an element of meaning essential to human understanding of the disease and its elusive character.

Back to the bioscience for now. Malaria in humans is the result of an infection by any one of five species of the protozoan parasite *Plasmodium*, the life cycles of which are quite similar. Infection begins when a female mosquito injects malaria *sporozoites* into a human's skin as she salivates while probing for blood vessel. These sporozoites soon glide through the skin and enter the bloodstream and then move into the victim's liver. There the protozoan undergoes a phase of asexual multiplication and, as *merozoites*, invade red blood cells to begin another stage of asexual reproduction that repeats itself almost indefinitely. Then some magic: The parasites transform to stages called *trophozoites* and *schizonts*, then back again to merozoites. Still with me? As parasites become abundant in the bloodstream, they reduce the efficiency of the infected red blood cells to carry oxygen, reduce the number of blood cells, and flood the bloodstream with their waste products. The body's reaction is to increase its temperature —*fever*—giving the affliction we call malaria. Or what most Ethiopians call *nidad* (being afire) or *woba* (fever). Both words connote malaria by describing an obvious symptom. In contrast, European terms for the disease (malaria or paludisme) come from their understanding of malaria's watery locations, referring to pre–germ theory ideas about the disease's most commonly understood wetland ecology. There is a bit of irony there.

Let's stick to the science for now. Meanwhile, the parasites swimming in the bloodstream periodically give rise to male and female cells (gametocytes) that circulate in the victim's blood. It is these sexual stages of the parasite that are critical for transmission from a mosquito when she bites. The gametocytes mate while in the mosquito's gut and then migrate to her salivary glands as she takes the blood. Fertilization occurs and the asexual offspring (sporozoites) wait patiently to be part of the fluid transferred in the bite of an unsuspecting human—or animal—host. This is an obviously complex transaction and torturous journey for those malaria parasites. The blood she takes nurtures her eggs with proteins and lipids, but also deposits the disease agents into that creature that provided—without their knowledge or consent—their blood. She drinks for her species/survival. And the complex, but incredibly persistent cycle of malaria continues. . . . No wonder it took Italy, France, Algeria, British India, Canada, America, and Britain over a century of medical science to figure this sequence out.[4] Mosquitoes and parasites did it as a matter of course, a matter of pursuing the inexorable needs of their own life cycles. Shall we blame them?

Telling the Story: A Matter of Scale

The chapters that follow are essays that move a story—malaria's human and natural ecology—through places and time. They present the malaria game (or dance) from different points of view and moments of discovery that took place in Europe and in Ethiopia. The goal here is to acknowledge the sometimes chaotic sequencing of knowledge, the human terms of belief in the laboratory and in malaria-prone communities.

The story told here opens in the eighteenth century and in a place remote from Europe and America, but central to the Nile Valley, in fact, near the river's source. It progresses to the nineteenth century when naturalists and medical science in Europe, America, and colonial laboratories groped for answers to the causes of the "fever" and why it occurred in certain places and at certain times and not others. The story then moves ahead to the optimism of the post–World War II world when people believed that travel to the moon and disease eradication were worthwhile goals (and possible ones), and sciences of parasitology, entomology, and economics could bring health and prosperity—eventually. Ethiopia and the ecology of "unstable" malaria is a good place to watch this dance, with its nuances of cultures and geography. Finally, the tale moves to a view from the mosquito herself, who cavorts, dodges, and adapts to the changing human setting around her. At times she seems under our control, but then she breaks loose and inflicts her toll on the elements that sought to control her and the disease she carried. The epilogue considers the moves of the dance and prospects for control, management, or even "eradication" as a final solution.

The choice here is to focus on Ethiopia: at the local scale of river valleys, of people's movements, ideas about illness, and the ecological interactions among mosquitoes, parasites, and people. Malaria makes its moves or dances its dance in local places. We grow dizzy even if we try to measure its effects and numbers globally or on a continental scale. The local nature of the story connects the ideas to real lives and real places that fall outside the numbers.

CHAPTER 1

Ethiopia's Malaria in the Age of "the Bark"

MALARIA IS A DISEASE, but it is also a story of affliction understood by residents of the Ethiopian highlands as an illness and cultural condition of the ecology, of the "other." In their view, outside agents cause the disease, and other people suffer (despite periodic epidemics). Malaria is thus understood not only as a disease of the body but as a belief that stems from the complex coevolution of human ecology and culture; it includes a vector (an anopheline mosquito), a parasite (*Plasmodium*—a protozoan), and beliefs about the nature of a human illness. As a whole, the story of malaria tells us of human settings of the disease, its shifting nature, and how it fits within the realm of human encounters and their biological world.

In February 1965, anthropologist Frederick Gamst was near his highland field site in northwest Ethiopia's Gondar region, near Lake Tana. An unexpected event along the main dirt road to the west aroused his curiosity:

> We saw a procession of about 30 people five kilometers east of Aykel in Chälga. These people carried broken gourds, or pottery, or pieces of *gwota* (earthen storage pots). Some of the people blew horns in the traditional style while others sung [*sic*] *hota* songs in Amharic. One man told us not to pass the procession or else misfortune would come to us. He said that he could not tell us the reason for this unique procession.
>
> Two days later a person who took part in this procession explained its meaning. He said that the procession originated in

Bäläsa [a highland zone to the east]. The various containers mentioned above held a spirit that caused malaria. This particular spirit was said to have killed many people in Dämbiya several years ago and very recently had killed people in Bäläsa.

The people in the procession said that this spirit was very heavy and thus weighed down upon them collectively. Also the malaria spirit did not want to leave the main road; therefore, the procession went along the road from Azazo to Mätäma [a town on the border with Sudan]. Starting in Bäläsa a group of people would carry the spirit in various containers to an area which was part of a day's walk to the west of their home. The inhabitant of an area thus reached would be compelled to carry the spirit the next day to an area further west, still in procession form. If the people did not carry the spirit further on, they would eventually be stricken with malaria, for this spirit was about to give birth to seven malaria children. Thus the malaria spirit finally reached Säkält and was on the way to Saraba when I last saw the procession. Eventually they hoped to have the spirit taken to Mätäma [i.e., the Sudan lowland border district]. Perhaps they intended to dump the malaria spirit over the Sudan border.[1]

This story illustrates the interaction of belief with the biomedicine of disease. The procession's vain effort to remove an illness (*woba*, or malaria) from one area to another place was a cultural response to a human disease that the culture of the highlands associated with a human ecology that was not their own. It belonged to the people and ecology—and the spirits—of the lowland border area. The procession was a response to an outbreak of malaria in 1964–1965 that evoked a painful memory of the earlier and deadly outbreaks of 1953 and 1958 that had killed hundreds of thousands.[2]

The disease's 1964 visit to the Lake Tana region was a reminder of highland people's view of human ecology as a set of oppositions that separated "self" from "other." The illness had been an unwelcome visitor, a traveling spirit rather than a permanent resident; something that in their view they could capture, remove—and depose. For them, malaria had a personality. Malaria's visit thus expressed not only physical illness but also a set of binary oppositions in human ecology in which malaria's cultural meaning revealed itself in their cultural and physical universe, which included, for example:

Lowland	Highland
malaria (endemic malaria)	malaria (unstable/periodic malaria)
Islam	Christianity
altitude (low)	altitude (high)
temperature (hot)	temperature (cool)
state (weak/distant)	state (strong/imperial)
soils (sandy or black)	soils (red loam)
lowland crops (sorghum/sesame)	highland crops (teff/barley/wheat/pulses)
language (Arabic/Nilotic)	language (Amharic/Tigrinya/Agaw)

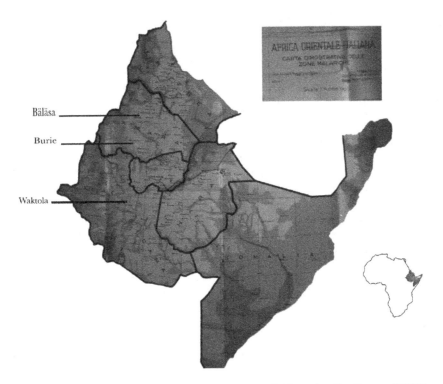

Figure 1.1. Italian map: Malaria ecologies, Horn of Africa (1938). (*Source:* WHO Archive, Geneva. Photo by author.)

Figure 1.2. Salubrious highlands. (*Source*: Watercolor by Johann Martin Bernatz [1844]. Photo by author.)

These contrasting concepts were fundamental to highlanders' ideas about who they were and how they separated their collective selves from others. Malaria was a part of the mentality of an ecology of personality, health, and disease. The stories in this chapter illustrate the interaction of belief with the biomedicine of disease. The 1964 procession was a vain effort to remove an illness of the body from one space to another and a flashback to earlier but still current beliefs about health and wellness. It was a cultural response to a human disease that people thought of as coming from a place outside, a human ecology that was not their own—a signifier of the peculiar nature of Ethiopia's unstable malaria.[3]

The Historical Ecology of Fever

Early church documents like chronicles or liturgical treatises occasionally refer in passing to *woba* or a febrile episode. If the wider world knew the disease as malaria, local people identified it by its symptoms: a headache, muscle pains, but especially fever. Malaria had different names in different parts of the highlands: In Shawa and around the country's center it was

Figure 1.3. Eighteenth-century reference to *woba* in Ge'ez-language church document. (*Source*: Hiob Ludolf Centre, Hamburg University.)[5]

woba; in the Lake Tana basin it was most often *nidad*, meaning "to light a fire," or "to be ignited"; for others it was *enqetqet*, "the shivering fever."[4]

In a clinical sense, fever was not always malaria, as blood smear microscopy or other more recent diagnostic tools show. In Ethiopia this fever, however, showed itself as a disquieting illness that not only sickened but frequently killed. One eighteenth-century parchment in the ancient Ge'ez language found in a Tigray church mentions *woba* in its text on the lives of saints; this is one of the earliest known references to malaria.

Though the exact date of this document (see fig. 1.3) remains obscure, it may be from about the same late eighteenth-century period in which James Bruce describes his experience with "ague." Bruce in about 1770 tells us about his own fever that he treated with "the bark" (powdered Peruvian cinchona bark, or quinine).[6]

Historical literature and indigenous cultural texts reveal local conceptions of malaria ecology as well as those conjured by visitors. Gamst's story of the antimalaria procession is a rare glimpse into what may have been a common response, one example of local strategy for expelling the unwelcome affliction, an attempt at sending it back to its own turf. Other stories and meanings from the nineteenth and early twentieth centuries are more elusive in their ideas about the place-related illness of malaria in particular, since that disease often appears in folk diagnosis through its bodily symptoms of fever, shivering, or as associated with culturally hot, lowland ecologies and cultural identities.[7] Calling a fever "malaria" in Ethiopia was a simple diagnosis when fever symptoms appeared in a certain place and in a certain season, as in Italy's Pontine marshes or in Boston's Back Bay.[8] It was a folk diagnosis of symptoms in a certain place and season that may well have been accurate but relied on local experience more than biological certainty.

We have disappointingly little direct historical or clinical evidence of malaria's presence and effect in the Ethiopian region in the nineteenth century. In the mid-nineteenth century, European biomedicine was still to be fully convinced about germ theory and had not directly identified the disease vector (the *anopheline* mosquito), the protozoan parasite (*plasmodium*), and the full ecology of malaria until the late nineteenth century, when the French army biologist Charles Louis Alphonse Laveran delivered his paper to that skeptical 1880 Paris conference, arguing for *Plasmodium* as the disease-causing agent.

Malaria fevers were endemic to many parts of lowland Ethiopia, and occasionally to mid-altitude areas as well. No one in Ethiopia or Europe associated the disease and its symptoms with the complex life cycle of mosquitoes/protozoans that would come to light by the mid-twentieth century. Yet in many world cultures, including Ethiopia, there was a strong appreciation of malaria's ecological epidemiology that related to its vector ecology (miasma—"bad air") and its symptoms of fever, lethargy, and sometimes death, even if they did not yet grasp the central role of the actual mosquito vector itself. After all, the bodily symptoms by themselves were misleading, since they also might indicate relapsing fever or typhus—both borne by ticks. Richard Burton, intrepid traveler in the 1850s, reported that Somalis in the southeast of the Horn of Africa actually believed that mosquitoes were the cause of the "fever." He dismissed that local belief as absurd (see below).

Early travelers to the Ethiopian region also identified *mala aria* by its location, its symptoms, and a treatment (quinine—the bark) that seemed to relieve or suppress fever symptoms.

Eventually seasonal change and levels of moisture in mosquito habitats, as much as bodily symptoms, became part of the diagnosis of malaria. Eighteenth-century Scottish traveler and naturalist James Bruce refers to a mortal fever that raged in the Lake Tana region seasonally from March to November, an epidemiology that rather clearly describes what we know as the unstable *P. falciparum* malaria, which also broke out in the twentieth century in 1953, 1958, 1964, and 1998.[9] The loquacious adventurer Bruce, who served as a type of imperial physician at the imperial court in Gondar in the 1770s, treated patients for seasonal fever symptoms using a type of *cinchona* bark extract—later known as quinine—that he may have had on his travel inland from the Red Sea coast and thence, years later, down the Nile to Cairo.[10]

Malaria had framed Ethiopia's human geography as it had done in other parts of the world. Fever—or rather its absence—determined where people lived, died, and set up community life, versus placeswhere they cautiously passed through. In Ethiopian people's local perception, languages, and folk remedies, malaria was a debilitating and sometimes deadly feature of those seasonally inundated edges of Lake Tana on its east and northeast shores, where the Blue Nile's flow emerged to begin its plunge to the lowland border zones and eventually to Cairo and the delta. The lakeside was a sickening place. In the 1630s, Emperor Susenyos had moved his lakeside capital to the higher site at Gondar, at least partially to avoid the lakeside area's fevers. That ecologically driven move initiated the Gondarine period (1730–1855), which historians think of as political history but that owes much to the malarial landscape's role. In those years, at the Nile's emergence site that was to become the town of Bahir Dar, there was an empty, seasonally muddy land; no urban population had settled in that site at the miasmic southeastern shore of the lake during or through the first two-thirds of the twentieth century. Malaria likely accounted for the lack of permanent settlement there until the foundation of Bahir Dar town in the early 1950s and its growth as a result of the Blue Nile dam construction in the early 1960s.[11]

But there is a much longer story of the encounter between humans, malaria, and cultural perceptions of the ecology of illness along Ethiopia's historical landscapes. In the nineteenth century, the increasing numbers of foreign travelers in Ethiopia noted malaria, or at least illness with its symptoms, as a hazard while crossing Ethiopia's hot Muslim camel pastoral zone on the east and Nilotic periphery on the west before reaching the cooler and salubrious safety of the Christian highlands. Henry Salt in the 1810s noted that fever was the cause of Ethiopians' "dread and horror of the coast," while in the 1840s Cornwallis Harris, who had traveled through the eastern lowlands to visit the highland capital of the Shawa region, quoted King Sahle Sellassie as warning him that the water of the "lowlands is putrid, and the air hot and unwholesome."[12] Harris himself contrasted the cool Christian highlands with the hot Islamic lowlands, waxing overenthusiastically about the wholesomeness of the "Abyssinian Alps" that looked to him "salubrious" compared to the feverish lowlands.

Malaria was at that time, however, not a precise diagnosis, nor was the mosquito vector understood in Ethiopia or in Europe until after 1880. Ferret and Galinier, two French travelers to Lake Tana in the early1840s, were, however, aware of the ecology of fever in the Lake Tana area and noted the deadly nature of the lowland ecology: "Toward the end of the rainy season the humid atmosphere and the soil, which is soaked and productive of a pernicious miasma, turn the country into a fatal region."[13] The same was true for all of Ethiopia's landscapes below two thousand meters elevation.

Malaria's history in and around the Ethiopian highlands reveals itself in a number of faces. Let's look at those pieces of the puzzle for Ethiopia. A review of those includes: symptoms and ecology, ideas about causation, therapeutics, and adoptions of local knowledge that sought to understand malaria and its human ecology of infection.

The Human Ecology of Fever

Europeans and other foreigners whose bodies encountered the subtropical deserts, coastal plains, and escarpments that surrounded Ethiopia's highlands freely expressed their trepidations about both the dangers and the realities of fever. Traveling in the mid-1850s, Henry Stern warned potential visitors to Ethiopia that "for more than six months of the year" the lowlands were "overspread by an atmosphere of fever."[14] For some, the fevers of travel offered the recurrent fever spikes that suggest the life cycle of the *P. vivax* malaria that they had encountered in India, Asia, or the Red Sea coast. Charles Johnston, writing in his journal in the 1840s, described a fear of recurrent fever symptoms:

> At the end of that time I became much alarmed at feeling the approach of symptoms threatening a return of the intermittent fever, from which I had suffered so much during the previous eight months. . . . The first decided recurrence of a fit of the intermittent fever, the paroxysms returning every other day, from which I had suffered so much in Bombay and Aden, came on during the afternoon of the day I returned from Ankobar [the royal capital situated at 2,700 meters altitude]. My illness, however, did not completely lay me up; for although on the day when the ague fits occurred it was with the greatest difficulty I could leave my bed.[15]

Walter Plowden, writing in the 1860s, evoked similar imagery of the returned febrile disquiet on the road while ascending on muleback to higher, healthier ground on the plateau:

> An hour took me up the hill, and again we breathed fresh air and cool breezes on the plains of *Tslala*; but as the sun grew warm, I felt the irritation of the blood, which I had began to experience from the previous evening; and after proceeding a short distance, I recognized, certainly without welcome, my old friend the *Massowah* [the Red Sea port] fever. Finding myself unable to sit the mule, after some efforts, I dismounted, and threw myself under a tree and, having sent two horsemen to search for water (which they happily found), I went through the exact process of the cold hot fit, with a force that rendered me almost delirious; I remained there till about three o'clock, when it left me.[16]

A decade later, Emilius A. De Cosson described a fever episode reminiscent of Plowden's:

> Of this day's ride I know little, for I was in a high fever and racked with rheumatic pains. The sun, too, was so hot that towards mid-day I had perforce to halt for an hour under a tree, as I nearly fainted in the saddle, and when we resumed the march my head swam so that I could scarcely distinguish objects clearly.[17]

Hunter Samuel Baker, also in the 1870s, wrote in his diary (later a book) about his wife's febrile episode as their caravan passed a flooded black cotton soil plain across the eastern Sudan frontier, heading on camelback toward the ecological refuge on top of the Ethiopian escarpment:

> No sooner had we arrived in the flooded country than my wife was seized with a sudden and severe attack, which necessitated a halt upon the march as she could no longer sit upon her camel.
>
> On the following morning, during the march, my wife had a renewal of fever. We had already passed a large village named *Abre* and the country was a forest of small trees, which, being in leaf, threw a delicious shade under a tree, upon a comfortable bed of dry sand, we were obliged to lay her for several hours, until the paroxysm passed, and she could remount her dromedary.[18]

Again, the symptoms of recurrent fever and aches would seem to mark *P. vivax* malaria, rather than the *P. falciparum* (the parasite that is most common and most deadly in Ethiopia).

Those symptoms that suggest malaria, however, often lead to post hoc diagnoses for which there are other possible causes (e.g., relapsing fever).[19] Relapsing fever in Ethiopia's lowlands was louse-borne and typical of the rainy season months of July and August rather than malaria's stalking ground of September through November. Mansfield Parkyns offered a helpful observation about different local diagnoses: "The natives seemed to have but one name for any fever caught in the jungle (*nedad*), whether it was common intermittent ague or the fearful bilious jungle fever."[20]

Here Parkyns's experience of travel and local life probably describes the difference between *P. vivax* infection (the "common intermittent ague") and the more deadly "fearful bilious jungle fever," which may have been either the deadly malaria *P. falciparum* or perhaps relapsing fever carried by either a tick or a louse bite. More recent medical evidence suggests that Europeans' and local traders' fever symptoms were malaria rather than relapsing fever. In Ethiopia's western lowlands, the louse vector is far more common (rather than the tick-borne variety). Lice took refuge in clothing and traveled between bodies huddled closely together during the height of the cold rainy season months of July and August when heavy rains washed away mosquito larvae. Malaria, by contrast, asserted its epidemiological force in late September through November once adult females emerged from their larva-stage habitat and began their blood meals.[21]

It may be, in fact, the seasonal ecology of the more severe fever type that helps us make informed retrospective estimations that the mosquito-borne malaria is the more likely one (i.e., that it is a companion primarily of the September-to-November wet season). These are what we might call ideas of ecology that connect illness to an environmental setting—perhaps not conclusive clinically but good and reasonable assumptions about malaria's past.

Notions of Causation

For mid-nineteenth-century Ethiopians and Europeans, fever, chills, and ecological settings marked malaria and its symptoms, not its then unknown biomedical causes (parasites, human hosts, and mosquito vectors). Local highland-habituated peoples and European visitors agreed on the ecological argument that connected lowlands and seasonal moisture with fevers that they often called malaria or its various local names (i.e., malaria transmission). Parkyns, traveling through the western lowlands of the Mareb River valley between what is now Eritrea and the Tigray regions of Ethiopia, tells his readers of the seasonal practice of the local farmers and traders to avoid the lowlands, but also of the temptations to take the risk/rewards of hunting, visiting neighboring market days, or observing exotic nature:

> In Abyssinia the *"quollas"* or deep valleys are the best places for natural history of all kinds. You must, however, be cautious not to descend into them at an unfavorable time, as in so doing there is great risk of being carried off by the fevers which prevail at some seasons of the year, and which are always highly dangerous, and often fatal.[22]

He clearly describes malaria's seasonality and its effect on travel:

> During the dry season, which is the time for hunting, when the heat and dryness of the atmosphere have dissipated the malaria, or rather removed its causes, the inhabitants of the neighboring provinces cross over to visit each other and to attend each other's markets.[23]

Others noted the lowland ecology as rich in vegetation and abounding in the noblest trees and plants of the tropics, while some pessimists disagreed, reporting that vegetation as "rank, malignant," or having "damp chill air crying fever, and a fetor of decayed vegetation smelling death."[24]

Overall, Parkyns offered the clearest assessment of the fever's seasonal character and the sense of ecology's role in malaria infection, albeit a false association with the direct cause. His thoughts seem to be more a summary of his observations of local knowledge and practice than a scientist's calculation:

> The season most to be dreaded is immediately after the rains (about September), and the two or three following months. The cause of the prevalence of malaria at this time of the year is evident: the streams, which have been flooded for a long distance on each side of their ordinary limits, retire, and leave pools and marshy spots full of quantities of putrefied vegetable matter, the exhalation from which are the cause of the evil. It is seldom that a traveler need find himself in these spots during the dangerous season. When he does so, it is more frequently from carelessness or foolhardiness than from necessity; for here in Abyssinia the valleys are so narrow that it seldom takes you more than a short day's journey to pass from one

village on the high ground to another on the opposite side. More-over, unless you are pressed for time, you need not travel in the bad season. It will be much better to rest for a few months in some comfortable place.[25]

Just as the associations between lowland ecology and fever based themselves more on local knowledge than on European nineteenth-century naturalist science, so did the understanding of links between the mosquito vector and fever.[26]

In the eastern region, where Somali people moved in lowland, malarial ecologies, Richard Burton griped about the annoying nature of flies and mosquitoes but mocked his Somali companions' conviction that the mosquito bites bring on "deadly fevers" and sneered that their "superstition" probably derived from their ecological observation that "mosquitoes and fevers become formidable about the same time."[27]

This Somali belief, was, of course, quite accurate and not superstition, but a fair ecological correlation by local folks who deduced that the season of the mosquito vector and the fever were cause and effect. It would take Western science fifty more years to confirm Somali observations about their disease world.

Mosquitoes, however, were indeed a nuisance, along with other insects that buzzed, swarmed, stung, and inflicted misery, particularly at night. Samuel Baker in the early 1870s described the insects that tormented Europeans, his Turkish wife, and local folks alike in his lowland camp:

> Not a man closed his eyes that night not that the dinner disagreed with them but the mosquitoes! Lying on the ground, the smoke of the fires did not protect us; we were beneath it, as were the mosquitoes likewise; in fact the fires added to our misery, as they brought new plagues in thousands of flying bugs, with beetles at all sizes and kinds; these, becoming stupefied in the smoke, tumbled clumsily upon me, entangling themselves in my long beard and whiskers, crawling over my body, down my neck, and my sleeping drawers, until I was swarming with them; the bugs upon being handled squashed like lumps of butter, and emitted a perfume that was unbearable. The night seemed endless.[28]

Ironically, those mosquitoes that most troubled travelers and local peoples were female mosquitoes of the *Culex species*, not the *Anopheles* malaria vector female. The latter female was she who "sang" quietly and sweetly as she often sought her blood meals from the more convenient ankles and feet rather than bothering to announce her presence by buzzing around her victims' ears.[29] Malaria, then, ironically, came almost silently among the nighttime miseries. After all, it was the ecology of altitude, season, and rainfall cycle that governed malaria's path; it was the human experiences of local travel, insect behavior, and vegetation's effects on mosquito population (such as maize effects on habitat nutrition—see chapter 5) that made the secret life of the lowland fevers.

If European travelers associated rotting vegetation and miasma with fevers, they had little or nothing to say about the agroecology of crops. Local ideas and expressions of illness, however, filtered into both belief and practice. In one expression of the *balamedhanit* or *dabtara* (medicinal healers), clerics in the central highland areas spoke of "Rahalo," a *zar* spirit who was responsible for malaria (or other fever symptoms). She was a traveling spirit (or one of several spirits) who may have been a version of what Gamst witnessed in 1964. That spirit had likes and dislikes, favored times and seasons of retreat, like mid-rainy season when she remained quiescent. In the local view, as an unwanted but foreign guest, the spirit could be captured and relocated to a neighboring home, village, or, even better, "escorted" across a national border. Her somewhat capricious behavior matched what malariologists would later call "unstable"—not an every-year infection.

Other local belief systems were more ecological than cosmological in their understanding of malaria and its febrile nature. Residents of the Lake Tana region over time had developed a folk taxonomy for their limited interaction with malaria, its causes, and its habits—fevers could have meaning in both the spiritual and the natural worlds. Highland residents' local knowledge of malaria's symptoms—and its cause, provenance, and prevention—was a product of their relatively limited exposure to it in some years and not others. Many of today's above-fifty generation recall the disease as exclusively a product of lowlands and marshes. *Nidad*, as they called malaria, was a peripheral disease of intermittent high fever that afflicted people who passed through the Abbay (Blue Nile) Valley or other lowland ecologies. Malaria afflicted especially hunters (*adagnoch*), political bandits (*shifta*), and Muslim merchants (*jabarti*), who, respectively, regularly traversed, escaped arrest by taking refuge in the lowlands, or carried trade goods between the major market centers on higher, safer ground in the empire. Rarely fatal, *nidad* fever affected travelers usually after their return home to the highlands above two thousand meters altitude, but the absence of the mosquito vector in the higher zones meant there was little if any transmission on the highlands themselves. Malaria was exotic, carried from the lowlands.

We now have a helpful contrast of views from clinics in Europe that had begun to grasp malaria's biology and had absorbed cultural understandings of ecology and treatment from various tropical experiences in Asia, Latin America, and Africa. The knowledges overlapped in time and space. Travelers often had their fever onset after the parasites had time in the human bloodstream to cycle and thrive. They often got malaria symptoms once they had returned home. The spirits were mobile at times of epidemics and appeared often well after the time of actual exposure.

Malaria, in the view of local healers, had personality. It was a disease of the "other" evoked by contact with the ecology of the lowlands or by malevolent spirits that travelers brought with them occasionally into highland space. This was the epidemic malaria that afflicted the highlands every decade or so. Collective action under spiritual guidance from specialist healers (the *zar*) could expel or exorcise malaria's spirits (as in 1953, 1958, or 1964). Another malaria form recognized locally was the onomatopoetic *enqetqet*, the "shivering fever" that sometimes affected those who slept outdoors alongside their herds at the edge of the low-lying, unpopulated wetlands (also see chapter 5 on maize and malaria).[30]

Overall, however, the historical malaria-free agroecology of highland areas above two thousand meters offered a reasonably stable balance between only mildly endemic malarial zones, a very limited mosquito population, the absence of a local human parasite reservoir, and temperatures below the point at which malaria parasites would not thrive. On cultural and biomedical terms, malaria belonged to the "other"—another place, another season.

Cultural Therapeutics

Within the literature of travel and its descriptions of ecology and local culture, we get a glimpse, but only a partial record, of the ways in which the travelers, their local companions, and the local population coped with the burden of the disease. The record contains an eclectic mix, sometimes weird, of pharmacopoeia, beliefs, and preventive practice that formed the human response to malaria as perplexed humans understood it well into the twentieth century and beyond. Malaria in Ethiopia and in the world of science was often a confused mix of beliefs.

External: Malaria Gestalt from the Outside

Foreign travelers and those who settled for longer periods in and around Ethiopia's highlands brought with them bodies of knowledge and preconceptions about health risks and amelioration of disease. In the mid-nineteenth century, germ theory in Europe was still on the diagnostic horizon, but drug therapies, such as quinine and antiviral vaccines, were emerging as a practice. Fevers, sometimes identified specifically as malaria, were a part of that groping for treatment without a real knowledge of what treatment worked, the cause of the disease, and why it appeared so seemingly randomly. On their travels to Ethiopia, European hunters, diplomats, and adventurers brought with them the stories of those whose travels preceded them. With these earlier published accounts of the hazards and their remedies, they brought the therapeutic medications of the day and preconceptions about the dangers of local ecology.

Some traveler remedies were amusing quackery, but others were tried-and-true treatments. Hardly systematic, their attempts at prevention—prophylaxis—were at an early stage of "hit-and-miss" medical science. Emilius A. De Cosson, who traveled in the early 1870s, used muriatic (hydrochloric) acid and quinine to treat himself after suffering the headaches, body aches, and fever experienced in travel in the arid expanses of the Sudan/Ethiopia border. His solution was quinine as prophylaxis but then a queer additional regimen of pharmaceuticals including Epsom salts, an "emetic" (a vomit-inducing drug), and occasional postfever doses of quinine "to prevent a relapse."[31]

De Cosson really just cobbled together "snake oil" elixers, but quinine was probably what actually had an effect on the fever. While neither Europeans nor local knowledge had grasped the role of the mosquito vector or the parasite in malaria infection, Dr. Parisis, a Greek physician at the court of Ethiopia's Emperor Yohannis IV (1868–1889), tells us that he used "sulphate of quinine both as a cure and as prophylaxis for malaria," giving 20–25 centigrams to the emperor and his courtiers every morning.[32] The emperor no doubt grimaced when he downed the bitter brew—he took his medicine, though we might well like to know what he thought about it.

Henry Stern also made quinine part of the regimen to gird his body and soul for the lowland climate and alien cultural setting of a sinful Sudan/Ethiopia border town (Metemma). For the ascetic Protestant Stern, the prescription was as much about escaping sin as the vagaries of fever: "Abstinence from almost all food, and eighty grains of quinine during four successive days, subdue the fever, and enabled me to leave that Lazar-house of vice, depravity, and crime."[33] Samuel Baker, in the end, skillfully relieved his wife's "paroxysms" of fever with a dose of quinine, a medicine that cured the symptoms during his lowland elephant hunting trips. No one quite understood how it worked.[34]

Mario Bourke, a hunter, came to northern Ethiopia in 1876 in search of the game that inhabited the lowland valleys of the Merab and Takazze Rivers—prime territory for malarial fevers. He had equipped himself well for the medical emergencies he expected to encounter and tells his readers that he included in his "most excellent medicine chest" the following: "A good quantity of quinine in two-grain pills, rhubarb pills, chlordane, a seductive solution of opium for diarrhea, Warburg's fever tincture, spermaceti ointment [whale oil], lint bandages, scissors, needles and silk for sewing up cuts . . ."[35]

Despite these preparations, Bourke, the lowland hunter (who foreshadowed Hemingway's ventures in East Africa in the 1930s), also reported that a mission station in the lowlands had lost one of their brethren to fever the previous fall. And then he himself fell ill to the fever. The best-laid plans . . .

Local Remedy

The illnesses they encountered often befuddled travelers, particularly in the lowland fever ecology where they sought prevention, treatment, and solutions of belief about the affliction. So they often related to the readers of their travel memoirs how they adapted local practice and experimented with hybrid solutions or tried their own ideas adapted to prepare themselves for the subtropical settings of Ethiopia's perplexing ecologies. These approaches were sometimes innocuous, sometimes alarmingly dangerous, sometimes just silly—or all of the above. Mansfield Parkyns, who feared the miasmas of nighttime, describes one of his strategies that may have drawn on local practice in the use of smoke to control insects:

> When I could get wood, I invariably lighted two large fires, and slept between them. This plan, though not very agreeable till you are used to it, is a capital preventive of disease; for during the day the sun's heat raises the moisture in steam, which, when the evening becomes cool, descends in the form of dew or fog, and in this form is one of the greatest helps to a fever. The heat you have around you answers the purpose of a local sun, and you are in no more danger than during the daytime. But when I say I lay between two fires, it must be understood that they were so close together that I was obliged to cover myself with a piece of hide or a coarse native woolen cloth, to prevent the sparks or embers, which might fly out, setting fire to my cotton clothes.[36]

To some this plan might seem a recognition of mosquitoes as harbingers of disease, but here Parkyns seemed to fear more the night's dampness than the mosquito threat.

Parkyns then described how local folk and those in his entourage of Ethiopians and local Arabs dealt with nighttime dangers, including the annoying insects that disturbed their sleep and, who, Valkyrie-like, seemed to ride the vapors of miasma:

> Another plan, which is always adopted by the natives, is not, I think, a bad one: Roll your head completely up in your cloth, which will then act as a respiratory. You may often see a [local person] lying asleep with the whole of his body uncovered, but his head and face completely concealed in many folds.[37]

Parkyns seems to endorse this remedy of wrapping the head to ward off breathing the nighttime dangers of fever, probably because it filtered the damp miasmic air and also kept buzzing insects from dive-bombing the ears of exhausted travelers trying to sleep. Ironically, of course, it allowed the silent mosquito malaria vector (female *An. arabiensis* or maybe *An. funestus*) to take their blood meals at their ease from the sleepers' ankles and feet—much easier targets with less chance of being flattened by a lucky blow of the sleeper's hand. But never mind the gruesome, bloody details.

Local people and their Ethiopian travel companions also kindly offered therapeutic prescriptions to their foreign guests to alleviate their fever symptoms of shivering and fever spikes. Charles Johnston describes in some detail his bodily response to a cold-shower therapy engineered by his well-meaning Ethiopian companions:

> The cure was to be effected by a kind of shower-bath, to which I was to submit, sitting down whilst the water was poured from a height upon my head, during the attack of the rigors which preceded the hot stage of the ague [fever] fit.
>
> The next day, accordingly, the water having been properly procured, on the first symptoms of the fit coming on, I sat down in the shade of a large *ankor* tree, a variety of the myrrh that grows at an elevation of seven thousand feet above the sea but yield no gum. Here, wrapt up in Abyssinia tobe [a cotton garment], which upon the first fall of the water I was to drop from my shoulders, I awaited the coming shower from above, for *Walderheros* [Wolde Qirqos, his

companion] had climbed into the tree, whilst some assistants lifted up to him the large jar which contained the water. The remedy, however, when it did come down, immediately, laid me full length upon the earth, for what with the collapse of the system attendant upon the cold stage and the cold falling water, it certainly cut short the fever, but nearly at the expense of my life, for even when I recovered from the first shock, and was taken back to my bed, I was delirious for several hours after, a circumstance that I have often had reason to be thankful for, had it not been a very usual symptom of my disease.[38]

In the Ethiopian Orthodox Christian context, the water was *tsabel* (blessed water), a therapy steeped in ritual and spirituality that could cure diseases of the human spirit and of the body. Strangely, this therapy was virtually the same (minus the theology) as what his contemporary Charles Darwin in 1850 had endured as a modern remedy in Malvern, England, for his own mysterious illnesses of the spirit.

This local prescription for a curative hydrotherapy failed (as it also had for Darwin at Malvern's modernist clinic). Its failure then led the weakening Johnston to accept an alternative treatment when he allowed himself during a cold stage of the fever to be "cupped in the Abyssinian manner," as a way of avoiding the inevitable hot fever spike in the afternoon.[39]

As Johnston searched for a way to cure or repress symptoms of his recurring fever, like other European travelers he also tried pharmacopoeia administered by his Ethiopian companions who had claimed experience in expiating fevers of the "other." In late summer his Ethiopian friend/companion Walda Gabriel concocted two kettles, one half filled with heated local beer (*talla*) and another with eggs, honey, and butter. Johnston does not mention an herbal ingredient, but Walter Plowden's somewhat later description of *gerowa*, an infusion made from dried and pounded leaves mixed with honey and melted butter, may have been similar. He described the result as a bitter drink, possibly a quinine-like tonic, and that the same bark when dried is also "recommended to be smoked in a hubble-bubble [water pipe]."[40] He does not tell us if he found that one helpful.

Medicine and illness were, it is clear from the documents of encounter, as much a matter of belief as of biology. Johnston tells us about the sincere efforts of his companion Walderheros (Walde Qirqos) to invoke a spiritual cure for his "harassing complaint" (i.e., recurring fever). It was apparently a fee-for-treatment service provided by local priest-healers:

These turbaned ministers of religion promised faithfully by prayers to cure me of my harassing complaint. I shook my head in a most scandalous manner, as I doubted the efficacy of their intercession quite as much as I did that of the devil worshippers, but gave them the salt [the payment] notwithstanding, and after a long blessing, which I thought would never have ended, these two holy men took their leave.[41]

Johnston's "devil worshippers" refers to Walde Qirqos's earlier summoning of a local healer/cleric (a *dabtara*) who probably sought to exorcise the *zar* spirit.

Practice of the Encounter: Dabtara Asres and the 1954 Malaria Epidemic

Stories that we distill from outsiders—travelers (Johnston and Parkyns); hunters (Baker); missionaries (Stern); or naturalists (Graham)—offer, at best, only platonic shadows of the local beliefs and practices at play as highlanders sought to cope with fever symptoms or disease in general. As outsiders, few travelers had any insights into deeper meanings of local medical practices. For their part, highland Ethiopians had ideas about wellness mediated by a set of those mystical clerics whose knowledge ranged from the experiential pharmacology of local plants to a rather solid sense of the spatial and seasonal ecology of malaria infection. That spiritual setting for illness saw malaria fevers as part of a world of male and female spirits that had personalities, capricious behaviors, and could travel along with particular cultural "otherness." [42] Frederick Gamst's comments on his 1964 procession cited earlier even indicated that malevolent malarial spirits preferred to travel on the main roads (dirt tracks in those days). Malarial fevers were foreign in both their home ecology (fevers were lowland-derived afflictions), seasons (wet season transition in September to early November), and in cultural milieu (Muslims, traders, political bandits, brigands).

We are fortunate to have at least one first-person testimony (i.e., an "emic" view) of the human ecology of malaria in a specific place during an epidemic outbreak in 1953. Dabtara Asres (1896–1985) was a cleric magician and healer interviewed intermittently over almost a decade by the French anthropologist Jacques Mercier. Mercier published his verbatim interview texts in 1988 in the volume *Asrès, le magician èthiopien: Souvenirs 1895– 1985*, translating Asres's autobiographical stories directly from Amharic to French. The text is a collection of stream-of-consciousness narratives rather than responses to specific questions. It is better that way. His life experiences were eclectic, to say the least. Asres was the son of an Orthodox cleric from Gondar north of Lake Tana, but he himself wandered, not aimlessly, between the life experiences of a frontier soldier, slave trader, healer, magician, interpreter of dreams, and, in 1953, malaria sufferer.

The loquacious Asres thus tells a great deal about the twentieth century, where epistemologies of malaria's bioscience overlapped with local practice. In the late summer of 1953, Asres found himself on the black cotton soil plain of Fogara, on the northeastern lakeshore of Lake Tana, northeast of the village of Sendegwe. He describes the local ecology that underlay the infection:

> The fevers caused misfortune in Foguera. This is extremely flat country and during this period nothing could grow there, not even eucalyptus like today. It was plain, absolutely bare. During the day the hyenas hid in Aferouannant, in Oudo, in Amora-Guedel, and in Dera, in the woods. They knew where to find carrion. At night, they did not stay in the hills; they descended on Foguera to feast on the dead cattle along the roads. . . .
>
> When the disease [malaria] came to the village of Sendegue, it left corpse upon corpse. The survivors fled. In their haste, they had time to take only a few precious possessions. Everywhere baskets of leather, hides which served as beds, belts, and bottles were abandoned. The hyenas that scraped the bones walked the grounds and entered the homes. They even pulled the carrions inside to devour them at their leisure. If by chance a calf or a cow came by, they attacked, biting its throat. Only the most curious [hyenas]came out during the day. The others stayed inside the houses and made them their homes.
>
> The epidemic made five victims at Sendegué and those who escaped came to find me at Kemkem [a nearby area].[43]

The village sought Asres's help in exorcising the disease's cause, and he drew upon his spiritual healing knowledge to offer an ecological diagnosis. But he divined that the time was not right:

> The *zars* (spirits of this fever). There are among thirty-seven or thirty-eight. They will not come out immediately. It is useless to try. I also called the spirit of the soil, the *Quolle.* Do not go! The

time for their [the zars] departure has not yet come. Do not make a move yet!

Frustrated, the villagers called another healer cleric, named Mekonnen and he prescribed a different approach:

"Prepare food for the ceremony. Bring your things and follow me!" he said. Men and women escorted him, clapping their hands and singing. He struck his stave [*merquz*] against the bell. He had long hair as if possessed. His assistant was carrying the cups needed to make the coffee. Upon approaching the village's fifteen or so houses, he told the peasants:

"Stay here and watch! I will go and make fumigations.[44] Once finished I will signal for you to come."

In his trance, Mekonnen the cleric then tried to enter a house to exorcise a hyena who had accompanied the spirit; Mekonnen broke his leg in the doorway. The hyena escaped. Women of the village took up their sticks and killed the hyena, but the priest/healer Mekonnen later died of his infected wound. The story in Asres's telling relates a strange reversal: hyenas living in humans' houses, women taking up weapons, and the healer losing his life. Asres's story offers both truth and fantasy in a curious mix. Asres the healer, however, clearly speaks of symptoms of malaria affliction:

At first the disease attacks the head/mind: the man complained his head hurts. It then spreads to the shoulders and then to the whole body. Some have diarrhea as clear as water, others have bloody diarrhea. Others it causes constipation. For those with bad diarrhea it causes pain and bleeding from the intestines. In any case, it always starts with a headache. First one person is struck down, then another. Five, six people in a house even. Not to mention neighbors. Everyone knows that it is the fever. That and the pox. It is extremely bad. Now they are gone. Formerly it was terrible. They could kill an entire family. Those infected would stay alive for only a few days.

Asres then tells us the epidemic's origin:

These diseases are *zar*. They are not demons. Demons cause the belly to swell, they gouge the eyes. They distort the body. Nothing like that happens with these diseases. They strike indiscriminately. There are many types of zars. The work of Sheikh Anbesso, Moras

of Abba, the Gragn [names of zar spirits] are different, they tell us.
They do not impose themselves on us and we do not impose our-
selves on them. We are under their whip. However, there are some
quolle that orders us to enter or leave. Leave! Why did you enter?
We are under their whip. [Asres speaks in the zar's voice now]: "We
make distinctions between us similar to those you make in society
when you say this one is a black person or this one is a king. We
invade a place like a buzzing swarm. If an unknown zar comes, he
joins us. We do not reject him; we live in friendship. *Zar* and *zar*,
demon and demon like each other. There is no conflict between
us. But a cleric can disturb [*irritate*] us as he fumigates us [using
smoke] from plants or the vomit and excrement of hyenas."

Asres, speaking again of the spirits, says:

They strike and move on. They do not spend the entire day with the
sick. If we spread ashes in the evening at the door where the fever
enters there will be footprints of children and adults in the morning. These are the prints of the zar. Ten today, tomorrow five, the
day after tomorrow seven. . . . They can increase to forty or fifty for
one house. In the evening they beat the drums to recall them. They
all come together deliriously throughout the night.

"They" in Asres's story are the malaria spirits—the malevolent spirits.

Malaria's historical landscape in Ethiopia has been a mix of beliefs, practice, affliction, and overlapping malaria meanings. The historical record, as
you can see, is spotty, but full of nuance. The next chapter moves more fully
into knowledges about malaria that lie behind current practice.

CHAPTER 2

Mindscapes of Malaria

Miasma in Two Worlds

> I also learnt that the cure was to be effected by a kind of shower-bath, to which I was to submit, sitting down whilst the water was poured from a height upon my head, during the attack of the rigors which preceded the hot stage of the ague fit [malaria fever].

> —Charles Johnston, Ankober, Ethiopia (1843)

> This in fact can be discovered and known in the power and the absolutely preeminent contribution of scholars like Golgi, Grassi, Celli, Bignami, Marchiafava, Dinisi, Bascianelli [who] gave the brilliant discoveries about malaria that have been given during the first years of our century. It is these together with Laveran, Ross, Koch and a few other lesser known foreigners who have posted the bases of all our present knowledge of malaria and raised malariology [to] the dignity of a new branch in the field of medical science.

> —Mario Giaquinto Mira, "La lotta antimalarica"

The Enlightenment in Two Worlds of Infection

FOR ETHIOPIA, DURING THE early modern world and the "long" nineteenth century, two perspectives on malaria merged, mingled, and overlapped. One view of the disease emerged from local cultural landscapes about bodily afflictions and their origins. Another set of ideas came from a modernizing world that, in the final decades of the 1800s, came to accept, albeit slowly, the ideas of germ theory—that contagious organisms entered the body from contact or via other carriers (vectors) and caused disease. Germ theory, interestingly, arrived in European medicine about the same time and in the same location as Europe's colonial expansion to the tropical worlds in Africa and Asia. Knowledge had embedded itself in experiences of culture, ideas about disease, and practices of healing for Europeans, Africans, and

Figure 2.1. Water therapy; priest and patient in Gojjam area (1975). (Photo courtesy of Dr. Jacques Mercier.)

those who traveled between those worlds. Earlier colonial expansions and mercantile trade had exposed parts of the world to one another, with new diseases presenting themselves to new audiences as Europeans explored new geographies of Africa, Asia, and the Caribbean.[1]

By the final decades of the nineteenth century in Europe and North America, germ theory dominated the understanding of disease origins and cure, and a sanitarian mind-set had become the vogue. The new view overturned previous understandings of malaria as a product of ecology—a miasmic watery landscape—in favor of drugs, a promise of vaccines, and new ideas about public health. Finally, the role of vectors (insects as vectors; rodents or snails as hosts)—intermediaries that transmitted bacteria, viruses, and filariae (worms) to humans—came to the fore. Malaria was a part of this web of intertwined knowledge and belief. Well into the twentieth century, malaria was a widely experienced global affliction, but there were decidedly persistent local beliefs about it in Europe and Africa.[2]

This chapter tells two stories about malaria that merged in Ethiopia and Italy in the years 1935–1941: intermingled tales of war, ecologies, and human struggles against a common disease that existed in Ethiopia as well as in Italy, places that in the classical era had both been part of a wider Roman trade network. In the nineteenth century, Ethiopia was a subtropical site of the European ardor for exploration and then an object for conquest. Ethiopia was the last of Africa to be the subject of imperial visions (Italy's, to be specific), and its highlands were home to ecologies that beckoned European imaginations about the land of a Christian ally (Prester John) and maybe even settlement by Europeans. Ethiopia promised to be to Italy what Algeria had been to the French in the 1840s and the Kenyan highlands had been to Britain in the 1920s.

Malaria was a part of that story, since it was an affliction that affected both worlds. Unlike the case for humid West Africa, Ethiopia's malaria was as deadly for local folks as it was for Europeans. Italy and Ethiopia had shared a common history with malaria from ancient times.

There were in certain areas "fatal zones," where population remained low, but there were also unpredictable, occasional tsunamis of the disease that once or twice a decade washed up from the lowlands to the highly populated highland places.

Worldwide awareness of a febrile disease occurring in the world's swampy areas appears in the oldest medical texts we have, including references on clay tablets and in papyrus scrolls and Sanskrit inscriptions. Not all fevers were malaria, but consistent fever/chills/headache outbreaks in certain ecologies led to folk diagnoses using the term "malaria" or other words that have come to be accepted descriptions of infections of *plasmodia* parasites in the blood. As in other cultures of health, Greek scribes, oral poets, and Ethiopian chronicles made ecological associations between marshy landscapes and fever with miasma and bad air (what Italians called *mal'aria*). Malaria or its telltale symptoms were in place in Greece and parts of Italian peninsular ecologies by the fifth century BC. Africa, however, seems to have been malaria's point of origin for what became the disease's global domain.[3]

The Western world, including Europe and North America, had had its own battles with malaria in places that seem now to be unlikely: in the British counties of Kent, Essex, and Cambridgeshire; in Boston, Massachusetts (exactly where I sit at this moment); in Minnesota's upper Mississippi Valley and the swamplands of the Carolinas. The world of the Mediterranean, along its southern and northern shores, was a prime fever zone, and included Rome's Pontine marshes south of the city—an area around what was to be the Fiumicino airport. Italy's malarial zones also included the hinterlands of the beaches at Anzio and Nettuno, where Allied armies landed in World War II and, in the northeast, the rich Po Valley, part of Venice's terra firma. Elsewhere, the classical Roman world included malarial zones as a part of ancient Rome's conquests in the west from the British Isles to Palestine in the Levantine. In each place the actors differed, even if the stage play was the same, with fevers, headaches, and shivering that was the result of the parasite that was *P. vivax* in Europe, and *P. falciparum* in much of Africa. The length of Italy's boot north to south had both vivax and falciparum, similar to the mix that had its way in Ethiopia.

If malaria had ancient origins in tropical Africa, it spread over time to the Mediterranean and other riverine civilizations in the Nile Valley, Mesopotamia, India, and South China.[4] The disease prevailed in certain ecologies of the eastern Mediterranean and beyond, and perhaps it is not surprising that the Macedonian Alexander the Great probably died of the disease as he returned from the eastern land margins of his conquests. One might consider that malaria's own empire has long surpassed that of Augustus Caesar and Tutankhamen (whose mummified body suggests that he had suffered from malaria) in both territory and space. In ecological (the Nile watershed and Red Sea) as well as cultural terms (Monophysitic Christianity and Islam), the Ethiopian region belonged to the world of Iskander (Alexander) in terms of trade and biotic connections. Malaria has had remarkable historical staying power: Once Rome lost its grasp on that territory, the world of Islam encompassed the malarial Mediterranean that also reached into Asia and the Arabian Peninsula, where the disease afflicted human hosts. *The Encyclopaedia of Islam* even goes so far as to suggest, with largely intuitive evidence, that the geography of the early expansion of Islam overlapped rather neatly with the geography of malaria.[5]

Malaria Speaks: Ethiopia's Languages of Malaria

We can begin to understand local ideas about malaria in the names Ethiopians have given to that periodic and unwelcome stranger that caused fever in lowland areas and with brief and deadly spillovers only sometimes into higher elevations. Malaria and linguistic references to its symptoms appear rather early in Ge'ez (or Ethiopic) texts under various names and events. The earliest I have found is an eighteenth-century document (in Ge'ez) held in the British Museum that refers to the devastating effect of *woba* (malaria) on an invading Persian army forced by a sudden outbreak of febrile illness—should we diagnose malaria?—to abandon its path toward an intended conquest of Ethiopia.[6]

Both early and modern Amharic dictionaries from the nineteenth and twentieth centuries also give us furtive glimpses of the vocabularies of malaria—a disease known for its symptoms of fever and shivering more than for a scientific biology of infection. Italian Oriental scholar Ignazio Guidi's *Vocabolario Amarico-Italiano* gives two words, *nedad* and *woba*, the first signifying malaria, but literally meaning the action of lighting a fire or something that is inflammable in the first case; and the second signifying a "tertiary fever, malaria," a "malarial country," or "something to become hot."[7] Ethiopian lexicographer Tesema Hapte Mikael's 1951 Amharic-Amharic dictionary gives the term *nadeda*, which it defines as "*woba* [malaria], *nidad* [malaria]; the shivering disease."[8] Thomas Kane's more recent and excellent Amharic-English dictionary quite effectively conveys the ambiguity of associations between heat/flame and malaria disease symptoms:

> näddädä to burn [vi], blaze, catch fire, kindle [vi], to scorch (heat: the soil; to get mad; to be resplendent, gleam (with decorations); to be afflicted with hunger, want, or cold, to be nipped (i.e., grain by frost); to be temporarily enfeebled, to lead, to goad, hurry up; to take away

> nädädä that which bursts into flame at the least contact with flame; very hot (lowland district)

> nedad yellow fever, malarial fever; warmth or heat from the fire

> nedadam fever ridden (place) location where malaria is prevalent[9]

For the alternative term *wäba*, he suggests more direct meanings:

> **wäba or woba** malarial mosquito, malaria; skinny Ar. wabā'
>
> **wäba tinign** mosquito [lit. malaria mosquito]
>
> **wäba bashita** malaria [lit. malaria disease]
>
> **wäba täwesak** malaria [lit. malaria epidemic]
>
> **becha wäba** yellow fever[10]

The dean of Amharic philology, Wolf Leslau, describes the word *wabā* as "plague, pestilence," or "disease, malaria."[11]

These meanings come from the active use of language to convey meaning of an experience, including symptoms of a disease, and incorporate knowledge of, for example, the malaria-mosquito connection and an insight into a body of local experience. A reading of a number of antiquarian and recent dictionaries of twelve Ethiopian languages (from four distinct language families) reveals a distinctive view of malaria in key words including *fever, malaria, mosquito, shivering, to catch fire,* and *heat.* The actual word *malaria* in common English usage comes from Latin and indicates the alleged cause of the disease, rather than its symptoms.[12]

So, within local languages in different large language families in the Ethiopian region (Semitic, Omotic, Nilotic, and Cushitic), borrowed words tell us in indirect ways about ideas and experiences. Ideas and suffering malaria's symptoms interacted within shared ecologies of local people's ideas of illness, its cause, and from whence they thought it came. Words for a disease, a plant, food, or location can come from experience or terms borrowed when cultures collide, or from ecological overlap and a language adopting a word that better fits a new phenomenon—such as a disease that appears only occasionally versus one that is an annual visitor. There appears to be considerable borrowing of terms referring to what appears to be malaria and its symptoms of seasonal fever, sometimes deadly, sometimes not. That borrowing may well have indicated malaria's behavior and movements in the past across Ethiopia's physical and ethnographic landscapes. If the Amharic language came to dominate Ethiopia's core highlands and the culture of the imperial state, it brought with it a sense of malaria (*woba* or *nidad*) as a rare invader from the lowland periphery that tended to inflict itself on "other" people.

Within the vocabularies of malaria in Ethiopia there is a subtle message about "place," which means certain sites or ecologies. This ecological idea about malaria actually resembles early ideas in Europe and the Americas where malaria was a disease of place and of season. Miasmatism, paludism, and swamp fever (or "ague," an older term from Middle English for fever/chills) are terms with variations used in English, Italian, and French or their local dialects.[13] The term "malaria" itself takes us back to the Latin of the Roman world and *mal'aria* (bad air) refers directly to ecology and an idea that malaria festered and then rose up from rotting vegetation in warm, watery sites. Ambivalence between vocabulary of a disease's ecology and its bodily symptoms has long been a sign of a society's recurring debate about the disease, its setting, its cause, and its signs. Language tells us quite a lot.

Chapter 1 has already described the era of quinine among European travelers and their attempts to adopt local understandings of fever used in local understandings of ecology of place and therapy. Europeans in West Africa or Ethiopia used quinine as antimalaria therapy, even if they had no clue why it reduced fever or cured its cause. Nonetheless, mid-nineteenth-century Ethiopians and Europeans had more or less agreed on malaria as a disease of ecology, an affliction of certain places. And part of the chess game was for players to stay as little time as possible in certain places on the board.

But dictionaries are rather static measures. One wonders how actual language use on the farm, in rural markets, and on urban streets came out when people talked about the "shivers" or the "heat" of fevers. We now know that, clinically, shivering was most likely a result of the release of *Plasmodium* parasites and their waste products as they burst from red blood cells. After all, to local folks, symptoms were the measure rather than microbiology. And all malaria is local.

A better bit of evidence of the subtle shifts of Ethiopian popular knowledge—what people know or what they think they know—is a newer translation of a venerable text called the *Ethiopian Orthodox Christian Book of Synksar*—a register of daily notations and biblical verses that offers a text in the old Ge'ez language with an Amharic translation in the opposite column. In the Ge'ez original text there is mention of prayer for a person "stung by the fever affliction" (*nidad*—which might be malaria). The original Ge'ez refers to *nidad*, while the modern translation into Amharic printed in the right-hand column sneaks in a modern bit of science—it translates that phrase as "stung by the *nidad* mosquito." In other words, the modern translator of this book of prayer, perhaps inadvertently, stuck in his own modern knowledge of the insect vector, unknown to the author who had composed the original text. That new addition misleads the reader and actually distorts the historical body of malaria knowledge.[14]

For Ethiopians, or for their neighbors in that region, up to the first half of the twentieth century there was little recourse but to accept an ecological point of view: Don't live in a malarial zone if you can help it. Good advice. Malaria was a disease of ecology of a place (a lowland) or of a time (shortly after the rains or in an epidemic year). There were no clinics, no government offering health care, no quinine for local folks. The prescription was not a medicinal cure, but a behavior—to move away or to avoid such and such a place in that season. The ecology locally marked where and when people could live in a place. Yet, equally, there was the vocabulary of symptom—heat (fever), chills, and the physical signs of shivering (*enqetqet*)—that coexisted with words about ecology of place. All malaria was local—it was an understanding that was as much about place as it was about the body.

For Ethiopia's malaria spaces, seasons, and ideas about therapy, it is important to keep in mind that there was in 1900 (as in 1936 and 1950) no local health service whatsoever available to the afflicted—no clinics, no hospitals, no pharmacies. Health stations in rural places showed up only in the 1960s, and then only in large towns or mission stations (see chapter 5).[15] Travelers or foreign physicians at the imperial court had "the bark" but only in minuscule amounts for themselves, the emperor's court, or for a small entourage. Ethiopia's peoples had to rely on signs of the illness and negotiated with malaria with their feet: by seeking refuge in highland elevations or in seasonal moving; or, if they lived through enough malaria in childhood, a very few had some "acquired" immunity. Malaria in Ethiopia was truly a phenomenon of local ecology and human avoidance response.

Malaria chooses its landscapes; humans had to adjust to that higher power. The disease, in reality, determined the history of human settlement—where people lived and where they did not. A satellite image of rural landscapes in Ethiopia of 1850 or of 1950 would have shown a patchwork of settlement by soils and rainfall, but also of malaria risk. High modernism in Ethiopia of the 1980s and beyond has sought—and usually lost the bet—to move people into those zones to develop the economies and improve people's lives (see chapter 5). But an earlier adventure by Italy and Fascism's colonial hand foreshadowed that false optimism. If Ethiopia offers us one case of human response to malaria, Italy offers another of high modernism where the belief was in science and an overarching and overambitious state.

Malaria and Nation: Malaria, Italy, and Ethiopia's Colonial Episode

Fascist Italy's 1935–1936 invasion of Ethiopia was a signal event in malaria's history, and a flashpoint that illuminated malaria as a modern, global disease. In 1935, the Italian air force dropped mustard gas on rural villages and Ethiopia's retreating army. In May 1936, Italy's black-shirted soldiers marched triumphantly into Addis Ababa. Those Fascist zealots also arrived with colonial ambitions for extracting agricultural products (e.g., semolina wheat) to feed Italy, plans to settle Italian peasants on fertile soils in a subtropical ecology, and to develop tropical products like cotton, sesame, and coffee, resources found in European rivals' colonies in Africa, South Asia, the Caribbean, and the Pacific. Italy had to catch up, didn't it?

But malaria stood in the way. Italian colonial officials brought with them to their Ethiopian adventure the knowledge from their own experience with malaria as a barrier to economic development and rural settlement in key parts of their homeland. Italy, uniquely among colonial powers in Africa, knew malaria as a domestic, not just a colonial, obstacle. After all, national hero Giuseppe Garibaldi's beloved wife, Anita, had died tragically of malaria in 1849. Indeed, the serious study of malaria as a disease between 1890 and its invasion of Ethiopia in 1935 had marked Italy's as the premier malaria science in the world. In the so-called Rome School of Malariology were great names like Giovanni Grassi, Angelo Celli, Camillo Golgi, Ettore Marchiafava, and Amico Bignami, who had championed germ theory as the new foundation for understanding malaria. And they applied these ideas in the treatment of their Italian patients. For these scientists at the opening of the twentieth century, the new nation (Italy) needed to address malaria as a part of its own nationhood and malaria, its number one public health crisis.

In Rome, on the eastern/western bank of the Tiber River—ground zero for the study of this pernicious disease—were the hospitals of Santo Spirito (for men) and San Giovanni (for women). Those hospitals and the nearby University of Rome had a steady flow of fever patients right at hand: Rome's Pontine marshes and the Roman countryside near its ancient port of Ostia had seasonal malaria patients who bore three types of the malaria parasites, *P. vivax*, *P. falciparum*, and *P. malariae*. What's more, Italian science in 1898 boasted the first researchers to identify mosquitoes (specifically, the anopheline species) as the disease's vector—the carrier between humans.[16] Under the guidance of the Rome school, Rome's malaria laboratories studied feverish patients, mosquitoes, and parasites—and the use of quinine as a therapy—on a massive scale. Italy's brutal incursion into Ethiopia in 1935–1941 offered a by-product, an entirely new landscape for the applied study of malaria.

If Italian science offered new insights from its own fields, it also indirectly laid bare malaria's damned persistence. In the 1780s Pope Pius VI had tried an ecological strategy for his home country by draining the miasmic Pontine swamps south of Rome but failed to control malaria. Again, as part of asserting Italy's new nationhood, between 1900 and 1914 scientists, with the blessing of King Victor Emmanuel's government, had been the first to apply Professor Giovanni Grassi's ideas to real-world human infections and confirmed anopheline mosquitoes—and not "bad air"—as malaria's carrier. He did that in 1899, first by importing infected mosquitoes into hospital rooms to meet "volunteers," who then had the predictable misfortune of contracting malaria. In that same year, Grassi and his fellow scientists tried the experiment in the field using metal window and door screens to protect some subjects from nighttime bites from *Anopheline labranciae* mosquitoes, Italy's most potent vector. Those screens, which Grassi called "mechanical prophylaxis," seemed to protect some volunteers but not others. In the experiment, those subjects protected from mosquitoes escaped infection: Of the 112 screen-protected subjects, only 5 contracted malaria. By contrast, all 415 subjects left exposed to mosquitoes came down with fevers, headaches, chills—all of malaria's symptoms. They got malaria. Voilà. Mosquitoes were the key link.

Grassi tried his field study again in 1901 near Rome's ancient port city of Ostia, this time with massive quinine therapy (aka "the bark") as a preventive—prophylaxis—on farmer patient "volunteers," some of whom received quinine and some who did not. The results of the "chemical prophylaxis" showed that quinine seemed to be every bit as effective as the antimosquito window/door screens. Of the 293 Ostia farmers who took daily quinine tablets, only 54 ended up with malaria, and then in its mild form. *Ecco!* Here now at least was a silver bullet that seemed to confirm the single-stranded strategy that would dominate malaria control for the next century.[17] Ecology had lost its place in the minds of European (and then American) antimalaria campaigns.

With these results and hopes for a solution in mind, the Italian health authorities between 1900 and 1914 began an ambitious quinine therapy program in malarial zones. Patients who came to clinics with fever symptoms, regardless of the cause, received quinine without cost, an effective treatment even if quinine had frequent side effects of diarrhea, vomiting, and nausea. Death rates from malaria dropped. Wonderful news—but with a caveat. Yes, mortality rates dropped remarkably. But strangely, overall rates of infection—parasites in the blood—did not. Even if their symptoms had receded, farmers' bodies were still full of parasites ready to infect others via abundant local mosquitoes. Malaria continued to afflict rural folk, but they at least stayed alive. But here was the rub, and a warning for future campaigns: Sick Italian farmers and farm laborers were willing to take drugs for symptoms, but resisted taking prophylaxis (quinine as preventive), the solution that Professor Grassi had envisioned. Farm household members took quinine from the traveling *cursori* (mobile antimalaria vehicles) when fevers and headaches came, and stopped when they felt better—a problem that public health folks call "defaulting." Meanwhile, those "cured" rural folks served as parasite reservoirs for the mosquitoes who sought blood meals and then nourished their eggs with Italian blood.[18] The Italian experiment had, in fact, saved some lives, but postponed any hope of eradicating the disease, and mosquitoes still roamed the marshes. Did this experiment in public health in Italy foreshadow antimalaria successes and failures in Ethiopia two decades later?

And then, in 1914, the chaos and disturbance of war intervened. Italy entered World War I's horrific trench warfare with the invasion of Austro-Hungarian troops in May 1915, and the movement of parasite-infected troops and civilians along the battle lines across the Alps and down into the malarial Po Valley. The resulting collapse of the health station network in rural areas spelled an end to quinine funding and delivery. In 1917–1918, the final year of that bloody war, Italy endured its worst malaria epidemic of the twentieth century. In addition to being the war to end all wars, World War I showed the failure of quinine as a silver bullet and, to the Rome school, proved that both peace (or at least stability) and an attack on the mosquito vector were necessary for controlling malaria.[19] But Italy's antimalaria failure and postwar political chaos merged malaria into policy and action with an emerging Fascist idea of reclamation of a Roman world and empire in both Ethiopia and Italy.

Benito Mussolini came to power in the weak, angry Italy in 1922, and by 1928 his government had sharpened an imperial ambition (first in Libya in 1911–1931) that later included Ethiopia. In 1925 Fascist health policy and faulty health science brought the irony of prescribing the use in rural kitchens of a salt laced with mercury, a silver-bullet solution to antimalaria research therapy in Apulia and Tuscany and distributed to World War I veterans working on land-reclamation projects there to compare its effects to quinine therapy and as prophylaxis.[20] The Italian use of poor, rural subjects in malaria research had elements that actually preceded the infamous Tuskegee syphilis experiments in the United States that began in the 1930s. But in the Italian malaria case, the government's approval of it foreshadowed the brutality of the 1935 invasion of Ethiopia with modern artillery, poison gas, and suppression of Ethiopian resistance in the following six years. Italy's invasion there brought the kind of social/political disturbance on which malaria thrived.

Ethiopia was quite different. Two points are worth making that had telling effects on Ethiopia's later experience. First, malaria in Italy was not yet a vanquished enemy—just one that had made a strategic withdrawal. Far from being checkmated by bioscience, malaria was simply beginning a new dance. When World War II fighting surged over the Pontine marshes in 1943 and the deposed Mussolini fled north behind German lines, the German army's scorched-earth tactics to the south of Rome included deliberately destroying the antimalaria water control structures. It was quite effective *environmental warfare*, though it only delayed the overall German defeat.[21] In short order, as the German army retreated from central Italy, malaria, which had been biding its time, roared back to occupy those disturbed spaces, and Italian efforts had seemed for naught.

Second, Italy's program in the Fascist years before their invasion of Ethiopia had come to focus on an entomological approach to malaria. Italian policy had, in the end, begun to focus on the entomology and ecology of malaria—to create barriers against mosquitoes, to drain the marshes, and to spray an expensive toxic insecticide, Paris Green (a very poisonous *green* copper and arsenic compound), in the swampy mosquito habitat.[22]

When in 1935–1936 the Italian Fascist state arrived in Ethiopia after the victorious army, it brought to the fray that same approach. First, they sought to identify malarial areas that had economic and strategic potential, then their field scientists moved to understand the exact mosquito types that might threaten the colonial settlers, soldiers, and administrators and then the local workers. The antimalaria goal, of course, was pragmatic rather than humanitarian.

When all was said and done, Italy's antimalaria muscle and experience were never brought to bear. Fascist Italy did, however, bring insights into the complexity of Ethiopia's setting as malarial science confronted Ethiopians' own ideas of practice and understanding of the scourge.

What was Italy's own view of its short Ethiopian colonial episode and malaria? In December 1940, the energetic Italian entomologist Prof. Augusto Corradetti published an article in the Italian journal *Rivista di Biologia Coloniale* that summarized four decades of field studies on Ethiopia's anopheline mosquitoes.[23] The journal's cover image was of a man in Arab dress on camelback and the inscription "incontro al sole" (toward the sun), evoking Italy's images and imagination of their short-lived colonial venture into Italian East Africa and Libya. There was some value and ambition there about malaria science, even within the tragedy of colonial violence.

Given that Italy's African colonies were to end the next year with Haile Sellassie's march back into Addis Ababa at the head of a British Ethio-Sudanese army, Professor Corradetti's 1940 summary of the state of mosquito research was a significant benchmark. It reflected two points of departure in Ethiopia's malaria history: First, it reveled in Italy's global

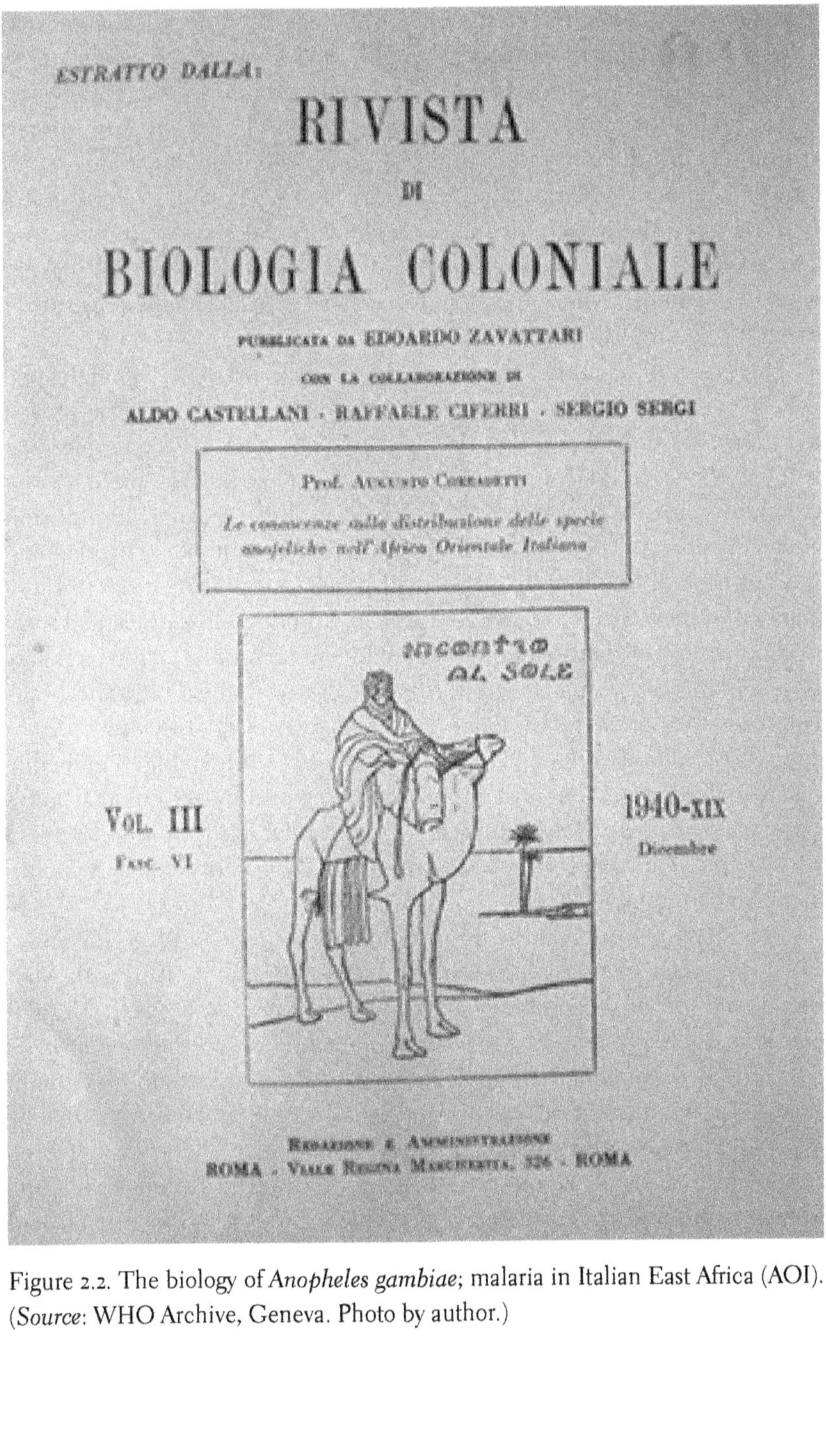

ESTRATTO DALLA:

RIVISTA

DI

BIOLOGIA COLONIALE

PUBBLICATA DA EDOARDO ZAVATTARI

CON LA COLLABORAZIONE DI

ALDO CASTELLANI · RAFFAELE CIFERRI · SERGIO SERGI

Prof. AUGUSTO CORRADETTI

Le conoscenze sulla distribuzione delle specie
anofeliche nell'Africa Orientale Italiana

INCONTRO
AL SOLE

VOL. III

FASC. VI

1940-XIX

Dicembre

REDAZIONE E AMMINISTRAZIONE
ROMA · VIALE REGINA MARGHERITA, 326 · ROMA

Figure 2.2. The biology of *Anopheles gambiae*; malaria in Italian East Africa (AOI). (*Source*: WHO Archive, Geneva. Photo by author.)

leadership in malaria research and its four successful decades of antimalaria work in Italy. Italian malariologists had advanced Laveran's 1877
evidence on *Plasmodium* as malaria's disease agent into an 1890 global consensus about a protozoan parasite that invaded and multiplied in human
red blood cells. Italian scholarship had gone on to explain various stages
and four types of malaria parasites (*P. falciparum, P. vivax, P. ovale,* and
P. malariae). The apogee of Italian leadership had come with Giovanni
Grassi's *Studi di uno Zoologo sulla Malaria* in 1900—a truly pathbreaking
work that, with his colleagues in Italy, laid malaria transmission directly at
the feet—so to speak—of mosquitoes.[24]

This is a story, of course, about malaria and Ethiopia. This part of the tale
is to view malaria as a backdrop to a larger political movement that proved
to be a precursor to World War II: Italy's invasion of an African empire that
had bravely staved off European colonization in 1896, but now faced a more
determined foe in Fascist Italy. Il Duce Mussolini's goal for the invasion
and occupation of Ethiopia that began in 1935 had a number of overlapping
motives beyond restoring Italy's national pride: One objective was to win a
colony as a market and a source for extracting products like wheat that were
in short supply in Italy. What Italian policy folks amusingly called a policy
of *panificazione*, we might awkwardly translate as Italy's "breadification."
Potential wheatlands in Ethiopia happened to overlap with the 1,800- to
2,100-meter altitude range that also included areas vulnerable to epidemic
malaria. The hunger for bread wheat and pasta wheat moved colonial ambitions to a new phase. Italy saw the greatest potential in sites within Ethiopia
that also presented the greatest possibility for malaria transmission. A conundrum and an irony.

Next, Italy saw highland Ethiopia as a neo-Europe, a place for settlement in certain of its favorite ecologies, allowing Italy to deal with what
they perceived as the problem of overpopulation in Italy itself.[25] Italian
planners wanted to address the wheat/population issue simultaneously by
resettling certain Italian villagers in parts of Ethiopia where they could
produce agricultural goods needed by the industrializing Italian economy.
Fascist policy called this "colonizzazione demografica," the conscious idea
of moving masses of people off the Italian homeland to relieve population pressure and to populate an empire in a way that the French had
done in Algeria (but not in tropical Africa) and the British had done in
Kenya (but not Uganda). The pilot projects were in areas of significant potential for wheat but, unfortunately, in their view also in ecological areas
vulnerable to outbreaks of malaria—sites that included classic highland
soils at Bishoftu (Debre Zeit); Holeta (west of Addis Ababa); Dabat (near

Gondar); Arsi (the SIMBA project for wheat); and Guder (for fruits and vegetables, including wine grapes). They gave these sites optimistic names like Bari d'Etiopia and Romagna d'Etiopia.[26] "La lotta malarica" therefore comingled malaria control with hopes for an agricultural transformation—*paneficazione*—that would underwrite Italy's African ambitions that drew inspiration from the British (Kenya), French (Algeria), and Belgian (Congo) models.

Italy's antimalaria program in Ethiopia envisioned four transformations in agroecology of control and economic extraction:

1. Settler immigration to ideal areas for Italian peasant settlement
2. Integration of colonial agriculture and agro-industry
3. Movement of a commercial agriculture economy to Rift Valley and Awash Valley zones for sugar, cotton, and oilseed ventures
4. Resettling of pastoral populations away from high-potential agricultural areas to more marginal, peripheral ecologies

Ultimately, these plans ran afoul of successful local resistance, economic failures, and the persistence of malarial settings that affected both Italian peasant settlements and highland Ethiopian workers.[27]

In some other sites that promised exportable commodities like coffee and citrus around Jimma, the Rift Valley, and Lake Tana, malaria was also a potential problem for Italian settlers and officials.[28] The seasonality and ecology of subtropical areas like the Ethiopian highlands coincided with ecological settings where anopheline mosquitoes cavorted quite scandalously. Italian occupying forces quickly relearned the lesson from their own national experience that all malaria is local. High modernism had worked in specific areas of Italy but was less promising for Ethiopia's varied ecologies. The primary Italian malaria vector mosquito was *An. labranciae*, one that liked brackish pools and outcompeted other mosquito types in the swamps south of Rome and the Po Valley, a habitat quite unlike the Ethiopian setting. Ethiopia's anopheline mosquito types were quite different from those Italian medical scientists had dealt with in Rome's Pontine marshes. Ethiopia was host to several malaria mosquito vectors—*An. gambiae, An. Nili, An. funestus,* and *An. coustani*—that had quite different habits of feeding times and breeding, and that could take egg-nourishing blood meals from animals as well as humans. These vectors also contained genetic variations that would not become known by Western science for another three decades. Ethiopian malaria was quite different.

Curiously, Italian policy in its homeland malarial zones had been to clear marshes, build local clinics, and provide drug therapy (quinine) to local farmer populations en masse. In Ethiopia the goal was not to promote the general health of the rural population, but to protect its Black Shirt officials, colonial soldiers, and the handfuls of Italian peasant farmers that the Italian government optimistically placed in those carefully selected sites. These project sites were for the Opera Nazionale Combattenti (ONC), veterans of World War I who had been the shock troops of Italy's own antimalaria settlement on the Pontine marshes.

The Fascist government's efforts at malaria control in Italy overlapped with its efforts that imagined Ethiopia as a safe, healthy colonial economy that was a part of Africa Orientale Italiana (AOI, Italian East Africa). In 1930s Italy, Mussolini's government—despite the short-lived mercury-salt debacle—had actually proved quite successful in using screened housing, quinine programs, and rural education to control malaria in its own backyards around Rome and the Po Valley. In fact, in 1939, the Italian government announced "victory over malaria" at a national exhibition at their program headquarters south of Rome. But at best, it was a Pyrrhic victory.

Yet, and almost inexplicably, the Italian colonial authorities also approved world-class malaria research that saw itself as a project of science (i.e., a massive effort to understand malaria and the broader ecology of key regions of their new AOI colony). After the Italian army's triumphal 1936 march into Addis Ababa, Italian research teams spread out across Ethiopia's key economic zones to assess malaria potential and other features of the local ecologies. Can we, in retrospect, reconcile the harsh nature of Black Shirt colonization of Ethiopians with the curiosity and motivations of malariologists?

Malaria research in the East African colony had to begin anew. Italy's self-confident malaria scientists had to quickly come to grips with Ethiopia's malaria ecology, which was very different from what they had tussled with in the Po Valley and the Pontine marshes. Step one was to identify the menagerie of anopheline mosquitoes that had evolved in Ethiopia's microecologies at different elevations and in puddles, hoofprints, and at lakesides and along river edges. Ethiopia's landscape ecologies looked and behaved nothing like the riverine flatlands of Italy. Between 1936 and 1940, Italian entomologists published eighteen separate studies of anopheline mosquito populations in various parts of Italian Ethiopia, many of which were conducted by teams led by the energetic Prof. Augusto Corradetti. Though Corradetti had signed the required Fascist membership papers in order to obtain permission to do his malaria fieldwork in Ethiopia, he later spent much of his postwar career in a distinguished career in malariology until his death in 1986.

The most ambitious and visible part of this plan was the ONC, the center point of the *colonizzazione demografica* (demographic colonization), which planned and began building villages at selected sites on Ethiopia's highlands that could absorb Italian peasants to produce wheat, fruits, and vegetables commercially.[29]

These technical malaria reports, published in Italian professional journals throughout Italy's forcible occupation of Ethiopia. displayed a rather remarkable awareness of malaria as a concern for humanity, even if the reports more or less acknowledged the economic value to the colony of malaria control, or, in the words of Prof. Mario Giaquinto Mira, "La lotta antimalarica" (the struggle against malaria). Giaquinto Mira's 1940 article appeared (in Italian) in an otherwise propaganda publication about the "civil organization" of the East African colony.[30] His vision for antimalaria efforts in Ethiopia bore some resemblance to Italy's own "heroic" program that involved placing peasants on *bonifica*, or reclaimed marginal lands. Giaquinto Mira's vision of Ethiopia's struggle began with a depiction of malaria as a global issue.

This grand Italian program of "la lotta antimalarica" in its East African empire also had in mind a global landscape that placed Italy's achievements in medical science and public health in the international context of colonial science. To Giaquinto Mira, Italy's efforts allowed it to stand above others and kept it on a global stage that included the Rockefeller Foundation's malaria work in Latin America (and after the war in Sardinia). He thus saw common ground with British, French, and German colonial medicine and their overall science:

> In this field as in so many others related, in human activity, Italy is known to be the world leader/teacher [*maestra*] such that in the glorious period of scientific study, tenacious, patient, and genial, it has opened the road with the discovery of parasites, vectors, and their complex biology, to the possibility of effectively combating malaria.[31]

Giaquinto Mira proudly asserts Italy's justifiable claim to leadership in global medical science, and also places malaria in the broader global struggle that included triumphs in Italy itself and in a colonial world. Italy's science also sought, in his view, a place with the colonial science of Britain (Ross in India); France (Laveran in Algeria); and Germany (Koch in Tanganyika). Assaulting malaria in Ethiopia was, for Italy, a pragmatic mission of economic development, but also a global crusade of science that placed Italian science on a par with the rest of Europe and America.

The antimalaria "struggle" was, it would seem, a center point of Fascist Italy's colonial ambition to capture Ethiopia's landscapes and boost its own national ego regarding conquering nature, resettling Italian peasants, and extracting agricultural rewards for the Italian economy. The "lotta antimalarica" set all three of these as goals, at least in theory.

Malaria control was as essential to the program of *sanitá* (public health), and the actual health (*salute*) of nonimmune agricultural workers, as it was for the Ethiopian laborers who constructed bridges and tunnels, traversed river gorges, excavated mountain passes, and crossed volcanic expanses. That work, however magnificent the engineering achievement, killed hundreds, if not thousands, of workers, who died not only from malaria, but as a result of falling rocks, dysentery, and exhaustion. Italian experience with the use of quinine to treat malaria in the Pontine marshes south of Rome and in Po Valley in the northeast was put into immediate effect for the workers, though not the surrounding rural communities. Yet colonial health officers soon realized that the vectors—the mosquitoes themselves—were very different from those of Italy. A 1938 survey of *anopheline* mosquito types showed at least 18 varieties of quite different potentials for spreading the agents of malaria; by the early 1970s that figure of recognized anopheline species had risen to 34 and to 41 by 1976. Among the Italian studies were elaborate trip reports and charts that showed spleen examinations and parasite load percentages, as well as the elevations at which various types of mosquitoes were found.

These maps of field studies offer rather convincing evidence of Italian science's efforts to understand the ecology of malaria, or at least the effects of altitude, temperature, and mosquito varieties.

The 1936–1941 Italian occupation changed the story, since Italy's occupying army, its plans for settling an immigrant population, and the health of a potential new workforce in agricultural projects required applying lessons from the Italians' own experiences with malaria. Italian colonial authorities commissioned an ecological, biological, and ethnographic survey of the Tana area as a potential site for economic development. The 1938–1939 study of Lake Tana was a tour de force of Italian scholarship in the sciences, but it was far too late in Italy's brief flashpoint of colonial history in Ethiopia to have an effect on policy or land use in the long run. The few copies of those volumes, printed on acid paper in soft cover, now languish in dusty archives with pages uncut. But those studies also bear remarkable testimony to an ecological snapshot of the coupled human/nature system as it existed

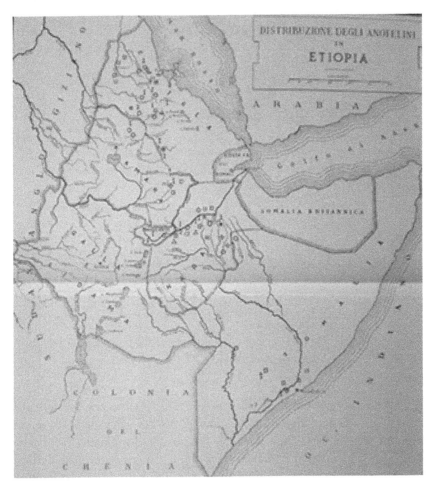

Figure 2.3. Distribution of the Anopheline in Ethiopia (ca. 1937). (*Source:* WHO Archive, Geneva. Photo by author.)[32]

just a year or so prior to the collapse of the Italian colonial apparatus and almost a decade and a half before the reconstituted Imperial Government of Ethiopia (IEG) in 1941 and its international sponsors had any effect on the ecological order of things.

A further fine example of Italians' remarkable capacity for ecological research and the grander plan for economic exploitation was the Missione di Studio Lago Tana, a project that eventually filled seven dense volumes of research on geology, domestic architecture, human body types, soil, hydrology, fish, and insects. Under the protective umbrella of colonial authority and the promise of a rich highland setting for economic exploitation, the

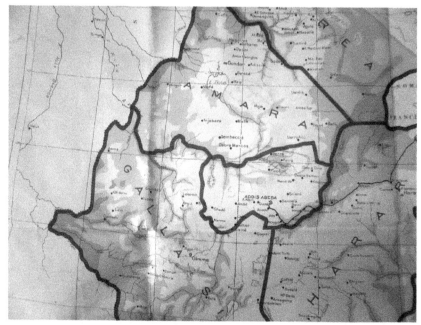

Figure 2.4. Italian map showing 1939 malarial zones (shaded) in the Blue Nile Basin of Italian East Africa. (*Source*: WHO Archive, Geneva. Photo by author.)

project set out teams of field-based scientists to do an impressive series of surveys of the Lake Tana Basin, which was the key watershed for the entire Nile system. What they quickly found was that Ethiopians historically had largely abandoned settlement near the lake because of its malarious character—marking the "fatal zone," as travelers had observed a century earlier. Malaria there was seasonal, but deadly. After only a few decades of trying to sustain a viable lakeside capital settlement at Gorgora, on the northwest corner of the lake, the site was abandoned in the seventeenth century for nonmalarial higher ground at Gondar.

When early nineteenth-century European observers arrived at the highlands from the Nile Valley or from the Red Sea coast, they witnessed a landscape at the ragged edge of the Little Ice Age. James Bruce's late eighteenth-century account tells us that snow was visible year-round on Ras Dashan, the highest peak. Later observers scoffed at what they took as exaggeration. By the middle of the nineteenth century, travelers' accounts also had described in reasonable detail the health conditions that faced them and, to a lesser degree, the populations they encountered in a range of environmental settings (see chapter 1). Travelers then described their journeys

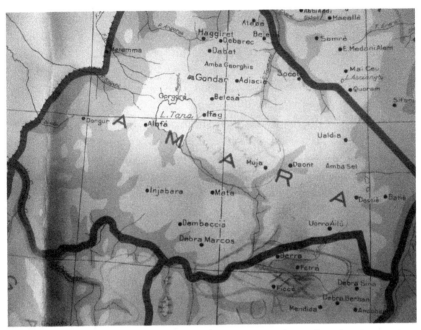

Figure 2.5. Italian map (1939) of malarious ares in the Blue Nile/Lake Tana region. (*Source*: WHO Archive, Geneva. Photo by author.)

and occasionally mused about Ethiopia's potential for European settlement, but more often showed interest in trade or in geopolitical alliances with the Ethiopian imperial state. European settlement was of little interest and global health issues were of little concern, but visitors were well aware of ague and dysentery as dangers to the traveler.

In October 1938, the Italian malaria officer posted in Gondar, serving as the Italian delegate to the International Office of Public Hygiene in Geneva—the League of Nations—presented a note from Professor Corradetti. In it Corradetti offered a description of the local mosquito varieties at five sites around the lake. He also included a drug therapy that reduced the amount of quinine prophylaxsis that he dispensed to his patients living below two thousand meters elevation to 600 milligrams three times a week on consecutive days for the months of August and September.[33]

In the 1941–1953 interregnum (called the British Military Administration), few, if any, records of infection rates or local outbreaks existed. One Italian physician who remained in the Jimma area through the Italian exodus recalled an outbreak in that area but could not offer any details, and health-care infrastructure—such as it was in Italian outposts—had collapsed. One

British malariologist, Sir Gordon Covell, who visited the area around the lake in 1952, found neither records nor coherent memories of outbreaks in the area.[34] Mosquitoes, however, romped, and the dynamics of malaria had free play with parasite-loaded lowlanders, people perhaps moving around less freely to vulnerable highlands since times were insecure, and holding malaria that still waited in local hot spots.

The Italian withdrawal and their summary route back to Addis Ababa in 1941 left only a skeleton of their tenuous five-year occupation. The road system whose construction had been so deadly to so many workers remained as silent testimony to Italian engineering genius. But it also had victimized force-marched parasite-loaded workers exposed to disease (malaria was one), injury, and death. In small towns or military outposts, such as Abalti on the southern cliffs of the Gibe Basin, there remained the stone house of the commandant and a single building with its stone arch where the few Italian and Eritrean enlisted soldiers had their quarters. The Italians had, sensibly, built their own quarters on the high point overlooking the Gibe Valley. There are other examples, too. Near the Blue Nile watershed divide at the old market town of Burie was a rough airport on the adjacent flat, low plain that did harbor malaria (and was thus deserted). In the town on the upper slope above the malarial plain an abandoned Italian half-track tank stood rusting for years in the school compound that had been their head-quarters. That position rather clearly showed that the malaria-aware Italians had chosen the flat, lowland field for their landing strip, but sensibly opted for the upland site five kilometers to the west for their military base. Sensible planning for the occupying forces.

The main, and impressive, remnant of Italy's occupation is still visible in neighborhoods of Addis Ababa and in the colonial cities of Jimma and Gondar, where the optimistic public architecture of art deco inspiration exudes a kind of modernism of style, if not substance. But none of those structures relay a story of public health or of successes in the effort to control malaria. In effect, the Italian occupying forces and state had done what Ethiopians had done for many generations. They retreated to highland areas above two thousand meters elevation, or moved by season, or simply accepted the risk of malaria infection.

La Lotta: The Legacy of Italy's Malarial Mission

When the Italian military forces retreated to Addis Ababa and then surrendered, they left behind a public health vacuum, a tabula rasa, where mosquitoes and *Plasmodium* parasites moved freely. The disruption of conquest and the resistance of rural areas to Italian administration left virtually no health-care delivery in Ethiopia. The collapse of Italy's own antimalaria efforts at home meant even less of an antimalaria legacy in their ex-colony—no file boxes of malaria research, no stores of quinine, only a few medical personnel in malarial areas.[35] In 1941–1942, the British Mobile Malaria Section of the East African Army Medical Corps made a survey of fifty sites (mainly in the Rift Valley and the Jimma area), reporting on "spleen rates" and altitudes taken from Italian malariologist Giaquinto Mira's 1942 report, from his apparent return to Ethiopia after the 1941 retreat. Overall, that cursory survey showed that malaria in Ethiopia was "predominantly an epidemic disease," though transmission seasons may lengthen in areas near rivers.[36] This finding was not really much of a revelation.

Human populations that celebrated the Italian departure and the return of the exiled emperor reentered a malaria landscape that was unchanged by five years of occupation. But the presence, defeat, and exit of the Italians left a war-weary human population to the mercy of the "disturbance" that malaria loved.

And the dance continued.

CHAPTER 3

Flight of the Valkyries

Malaria Ecology in the Headwaters of the Nile

> I found myself attacked with a slow fever, and, thinking it was
> the prelude of an ague, with which I was often tormented, I
> fell to taking bark.
>
> —James Bruce, *Travels to Discover the Source of the Nile*
> (1770)

> For two months I was sick. I felt, what? The smell of roasted
> coffee excited me but when I drank it, it tasted like bitter aloes.
> And now, what next? The spirit [zar] who made me sick or-
> dered me: Lie down on the right side! Turn to the east! I re-
> mained unconscious until the evening. . . .
>   A man as bright as the sun presented himself before me.
> He wore a garment that exposed his arms and chest. When I
> broke into a sweat and I vomited, I saw him appear majestic
> and bright. When we began moving, my soul emerged tame
> and weakened like a puppy. . . . All day my soul was traveling.
> The road stretched out indefinitely ahead of me, as straight
> and smooth as asphalt from west to east. Along its sides people
> were sitting in the sun. He did not salute them and they did
> not invite him to sit down. I followed him like a dog all day
> without any fatigue. In the evening we rested for a while in an
> unfamiliar place. Then he brought me back. As I entered the
> house I regained consciousness in my soul and body.
>
> —Asres's oral memoir of his malarial fever delirium (1953)

IN OLD NORSE LEGENDS, a Valkyrie is one of a host of female spirits
on horseback who fly over a battlefield and decide who will die and who
will survive. Malaria landscapes were a kind of battlefield with physical and
human contours moving and convulsing over time. Elements of that geography included biological players, but also spirits and ideas about illness. The
Blue Nile watershed in the late eighteenth century was a human ecology
that already bore the imprint of natural and spiritual forces. Most people

lived in the highlands, where their livelihoods of agriculture and local trade kept them out of reach of seasonal, occasional malaria. Only the very brave, bold, or foolish tempted fate by straying into lowland areas in the wrong season. The western lowlands leading to the Nile Valley were still terra incognita for most highlanders. And the usual written local sources of history, such as royal chronicles, hagiographies (lives of saints), or bits of correspondence, that survived in the nineteenth century involved spirituality or political legitimacy and had little to say about conditions of the physical world such as health or disease. Accounts do occasionally hint at a widespread famine (*erhab*) or disease (*talalafi besheta*), but those references are short with few details. The historical documents offer little that would satisfy a curious modern medical researcher. Let's focus on stories of malaria in one ecological landscape and tease out malaria's dance over time.

The Blue Nile has its actual, and symbolic, origins in a spring that emerges from swampy ground at the church of Saqala Mikael, a sacred site above ninety-five hundred meters elevation. From there the Nile waters begin a journey into and then out from a broad but quite shallow lake called Tana. As it leaves the lake's southeastern corner, the river gathers momentum in its descent, spills forty meters over the Blue Nile Falls, and plunges into a gorge deeper than the Grand Canyon that carries it in a wide arc, rushing in its deep, layered geological bed of stony red and blue rock layers to the southeast. The river then turns to the southwest, then west across the Sudan border. The riverbed then flattens into the Sudan Gezira plain and merges with its smaller cousin, the White Nile, at the "elephant's trunk," the city of Khartoum, and then goes on to water the glories of Egypt. That watershed collects its water and its momentum as it flows and defines a region, shapes a cultural identity, and frames a landscape for epidemic, episodic malaria.

The Blue Nile's turbulent, watery sweep has, over time, formed a unique ecology from the natural forces of nature: water, gravity, soils, and the imprint of human settlement, technology, and disease. Malaria has long lived there. And the shape of the land and the rhythms of the seasonal rainfall that washed soil down the Nile Valley formed the basis for one of the great moments of human history in Egypt and Nubia. For centuries, until the 1898 building of the first Aswan Dam, the Blue Nile (or Abbay, its Ethiopian name) and the Nile as a whole were little disturbed by the human actions of the Ethiopian farmers' scratch plow, their annual cereal cultivation, or the caravan traffic of mules, camels, or donkeys led by intrepid Muslim traders (*jabarti*). Rainfall on the earth's surface was a persistent watery pulse that took rich soils from the volcanic highlands and then steadily eroded them into the Nile flow as reddish, fertile silt—Ethiopia's true gift to the Nile's history.

Lake Tana is at the Blue Nile's first major outflow and is the heartbeat of the watershed, registering its seasonal pulse.[1] Lake Tana and its surrounding ecologies act as the font of the Blue Nile's waters, sitting at eighteen hundred meters above sea level from where it frames much of the watershed. The lake's geological origins were a volcanic blockage of the river's stream about two million years ago by a fifty-kilometer flow of molten basaltic rock that trapped water in the shallow lake bed. But the lake's water was a dynamic, changing flow and there is also evidence that the lake dried up 15,100 to 16,700 years ago and became a papyrus swamp; this was a major change in lake ecology and mosquito habitat. In this dry-phase historical era, geologists estimate that rainfall was less than 40 percent of what we suspect has been the pattern in the last two centuries. Then, about 14,750 years ago, water in the entire Nile system seems to have increased and the lake overflowed the volcanic blockage into the Blue Nile riverbed again. This process seems also to have happened in East Africa's White Nile (the Nile's junior partner) when Lake Victoria, also in a dry phase, filled to overflowing. The Nile system in its more recent geological history has regained its permanent flow, though still with the strong seasonal, pulsing rhythm that drives the Blue Nile watershed's ecology and gives us the modern ecology of malaria.[2]

In its recent history, including the twentieth century, the lake has been a shallow vessel, with a maximum depth of only fourteen meters (forty-two feet). Four small rivers drain into the lake, and the Blue Nile is the lake's only outflow. The lake's oxygen level is unusually high. Though the flow of water connects the river's water and the silt it carries through the lake and into the great Nile flow toward Khartoum, the lake's discharge into the gorge at the river's outflow makes up a total of 8 percent of the annual total into the overall watershed. The river's and lake's fish biodiversity results from the dramatic fall of forty meters at the Tisisat Falls ("Smoke of Fire") thirty kilometers downstream from the Blue Nile's outflow. Aquatic life—its fish species, plant life, and malarial habitat—around the edges of the lake itself developed its own historical character (its endemism) isolated from the wider Nile system. Fish cannot swim up the forty-meter falls or in the river's cascade down the deep gorge, and other aquatic organisms did not ascend the river to the lake. Twenty-eight fish species inhabit the lake, twenty-one of which are endemic. Seventeen of those species are large barbs, types unique to this place and its ecological cocktail.[3]

The eastern and southern shores of the lake display its malarial ecology in swamps dominated by papyrus, water lilies, and wet "black cotton" soils.

Large and sprawling ficus trees (*Warqa*) overhang the water's edge, providing nesting spots for African fish eagles, whose shrieks are among the first sounds of morning. Yet it is not the lake itself but its seasonally flooded edges that form its malaria setting. The lake itself offers few habitats for mosquitoes, but the surrounding wetlands and puddles formed by receding lake water in the early dry season provide a habitat for *An. pharoensis* and *An. funestus* mosquitoes. These mosquito types may be the primary vectors that have made the lake area historically a "fatal zone" along its edges.[4]

Malaria has long lived there. Through the millennia, the actions of Ethiopia's people in the highlands did little to alter these slow rhythms, compared to other, deeper effects of climate fluctuations like the Little Ice Age or the seasonal pulse of rainfall, surface water flow, and annual agriculture's scratching out human livelihoods on what European visitors called Ethiopia's "salubrious" highlands.[5]

The Blue Nile Basin, its geology and its geographies, shaped the cultures of several distinctive peoples. The ethnographic landscape included Christian highland farmers and aristocrats, Cushitic-speaking Agaw farmers, Muslim traders (who spoke Amharic), and Omotic-speaking Shinasha people. At lakeside there was a cultural and economic anomaly in the Wayto people, who made their livings from fishing, hippo hunting, and papyrus boat transport. Living at the lakeshore without agriculture, the Wayto adults had probably developed an acquired immunity to malaria, though having to bear the burden of high mortality of their infants—a heavy price to pay. These various local cultures traded places and bodies of knowledge on the local ecologies over time. The watershed is not a political boundary, but it has a political border, since an obscure 1902 treaty included an international border with Sudan, even though its waters and human ecology tend to ignore such ideas of politically generated boundaries. Within the change of seasons and human movements, malaria and other diseases danced in the ecologies of elevation, watery habitats, and human struggles to adapt to them. Epidemic, unstable malaria is one of the deadly disease outcomes of this human/natural ecology. So, too, were other enduring maladies of the watershed, exotic diseases of the subtropical world. We can count among these rinderpest, kala-azar, yellow fever, tick- and louse-based relapsing fever, and lower intestinal tract infections.

More than canaries in a coal mine, malaria infections were not a sign of a problem but were the problem itself. Malaria's fevers and human sufferers gave us the human landscape. In the longer-range past these were the drivers of where people in the watershed lived and when they could move about. The disease and the appearance of its vectors marked seasonal change almost as much as the rigors of the agricultural seasons of planting and harvest, though in the higher areas malaria was an unstable marker of time.

Malaria and these other diseases fundamentally molded human history in the watershed. But malaria in its epidemic forms was by far the most deadly local disease that shaped human movement. Malaria was the aggressor and humans responded defensively, moving to safe areas but periodically enduring epidemic outbreaks that swept from lowland hotspots to settled highland populations. Malaria's movement was the lead partner in the shifting dance rather than just an opponent sitting patiently on the other side of the chessboard.

Peoples of the Blue Nile watershed had, over time, married their annual cycles of growing food and their political life and religious ritual to the ecological setting. Environmentally, the Blue Nile watershed was the first part of the northern highlands to receive rains in the late spring/early summer months from the band of heavy rainbearing clouds arriving from the southwest, whereas shadows in the eastern hillsides often miss out on the rains so essential for agriculture. Trade between the region and the Nile and the Red Sea percolated across ecological zones that were malarial in relatively predictable seasonal rhythms. The lesson was to avoid lowlands in fever months. Malaria did, however, spill onto the highlands once or twice a decade (as best we can figure).

Malaria was an unwelcome but occasional visitor to the Blue Nile watershed's highland zones where people lived, and a more permanent resident of the lowland areas where only people on the political or ethnic margins dared to live. Malaria was, to some along the edges of ecological zones, a regular part of life. But it was a delicate balance of early death for infants and sometimes for adults, since Ethiopia's "unstable" malaria meant that virtually no highlanders developed acquired immunity. Those few lowland dwellers who survived to adulthood in the hot lowland zones on the west likely had built up an acquired immunity. Life was a risky business.

This chapter now addresses malaria's flight of the Valkyries, where the malarial spirits chose when and where to descend on their human victims. Maybe it is not surprising that highland farmers attributed malaria's whimsy to spirits rather than the actions of a lively, physical landscape.

Over time, the Blue Nile watershed's malarial paths have suddenly appeared in the lowlands, washed over the highlands, and then receded—a rhythm that has repeated over Ethiopia's *longue durée*, but suddenly and unpredictably. The patterns are predictably periodic, episodic, and maddingly erratic. Here I want to illustrate malaria's behavior by highlighting four episodes of the malaria/human dance in the Blue Nile watershed—with visible drama on a global and local scale. These episodes illustrate scales of place, time, and stage to display a rather astonishing complexity. The science and human frailties that appear in malaria's history sharpen our focus as we move toward the present.

Episode One: James Bruce's Lake (1770)

Unfortunately, for our glimpses of the Blue Nile's malarial history two centuries ago, we rely largely on outsiders. James Bruce's multivolume description of the region included bits of his daily life in the cooler highlands of the royal capital at Gondar, but also the adventurous, tall (six-foot-six) Scot's experience with travel from the Red Sea lowlands and into politically marginal areas where he could describe health conditions that do not appear in Ethiopia's own historical records.

Bruce, in his time, was a much celebrated and often reviled observer of the Nile Valley—strangely reviled by his own countrymen who disbelieved his egocentric stories. But his fame spread his views widely: Thomas Jefferson had a copy of Bruce's five volumes in his library at Monticello, and the Boston University Library's 1790 five-volume set had been a gift to the Boston Public Library from President John Adam's personal library in Quincy, Massachusetts.

Bruce had an enlightened Scot's skills at the medical arts, a naturalist's curiosity about geography, and an ego that kept him in close contact with local politics. He loomed large in his own legend.[6] Those qualities meant that his accounts describe his own maladies, such as observations of a smallpox outbreak, his recurrent fevers of "ague" (i.e., malaria), and his use of "the bark" (quinine) to relieve fever symptoms. He survived those bouts of fever. Did he have the more deadly *falciparum* malaria so common in Ethiopia, or was he lucky and contracted the less deadly *vivax* type that brought recurrent fevers but less often death? We will never know. His blood trail has long since dried, and we must guess that *P. falciparum* might have killed him, but *P. vivax* would have spared him.

We can, however, reasonably assume that Bruce's stories of life at Gondar and its ecological periphery around Lake Tana tell us about a disease of fever that must have seemed to Ethiopian elites (his friends and rivals) to be a normal—unremarkable—hazard of health. Malaria for Ethiopian highlanders, lowlanders, or Muslim merchants (who traveled between the two) was a normal force of nature and sometimes fatal, but it was a regular enough part of life for it to be unremarked upon in the local historical records.

Bruce shared his era's obsession, not with understanding malaria's deadly curse, but with finding the "source" of the Nile, which he believed lay in the watershed of Ethiopia's northern highlands. Ironically, it was probably a bout of malaria — or "ague," as he called it — that delayed Bruce's eventual visit to Gish Abbay, the site of the Nile's legendary source spring at Saqala. When Bruce finally arrived at that site on November 4, 1770, he described in some detail the location where two springs emerged from the ground at the foot of Gish Mountain, seventy miles south of the lake. He described his reaction when he came into sight of a rustic, newly built church and caught a glimpse of a marshy patch: "I came after this to the island of green turf, which was in the form of an altar . . . and I stood in rapture."[7] Bruce, in his own mind and in the stuff of local legend, had reached the source of the Nile (and the hydrological starting place of the watershed). Too high in altitude to support anopheline mosquito breeding or malaria transmission, the Gish Abbay site nonetheless marks a point of origin for the water that forms the central stream of the watershed as it gathers speed and makes a looping right-hand turn for four hundred kilometers into the full Nile Basin in Sudan. Overall, the Nile's flow has not changed much in the past two centuries. The waters in the river gather in June, surge until mid-September, and then slow to a trickle by December to the following June, when this climate event begins again. But the seasonality of the river also is driven by the temperature, the pooling, and the human movement that make malaria a part of the ecology of the watershed as a whole.

The ecology of the Blue Nile watershed in the 1770s that James Bruce witnessed was a force of nature in transition. Water and nighttime temperatures during those times were different than in modern periods. Climate science using soil profiles, lake sediment cores or ice cores, and glacier records on the landscape itself tell us that in 1770 parts of the planet were in a cooling period brought on by low solar radiation, changes in ocean temperature and circulation, and perhaps volcanic activity. That period began around 1550 when a "glacial expansion" — the Little Ice Age — took place and lasted until the mid-nineteenth century, including the period when Bruce traveled the Blue Nile region. If Bruce tells us that he saw snow on the top of the peak Ras Dashen we might believe him, even if that snow was not visible in the twentieth century. We might well read that as a sign of what the 2007 International Committee on Climate Change called glacial expansion.[8] Snowy icecaps in Bruce's time might mean that nighttime temperatures would have been lower: fewer mosquitoes, fewer and less potent falciparum parasites maturing in mosquito guts, and perhaps later dates of harvest for grain crops. But what were the broader effects on health, politics, and the overall human condition in local settings in Ethiopia? This is a tough question that is worth asking.

The end of the Little Ice Age, marked commonly at around 1850, coincided with climate change but also a rebirth of political centralization in Ethiopia. European arrivals into Ethiopia's malaria ecologies coincided with Europe's new colonial ambitions. We know few details of climate conditions, but we can surmise rising temperatures, especially at nighttime. Mosquitoes and European adventurers may both have found the warming climate to their liking. At least that is a possible hypothesis about historical climate change.

Resurgence of regional trade across ecologies of elevation may have slightly changed the geography for malaria transmission, even as political changes engulfed the Ethiopian region. We know something about political change in the mid-1800s; we know next to nothing about the social or ecological life of mosquitoes or malaria in the Blue Nile watershed. Travelers reported their fears of fever, but voices of the true status of malaria in Ethiopia were silent for decades.

The first genuine break in that silence came three decades into the twentieth century with the Italian invasion of Ethiopia in 1936 (see chapter 2). The Italian Fascist state sought to prepare Ethiopia as a colony of extraction for agriculture. During Italy's five-year occupation, its prodigious malaria science had a half decade to survey malarial landscapes and catalog mosquito species in the region. It offered a new perspective of the ecologies in which Italy sought to gauge how potential Italian settlers and local laborers could survive to feed Italy and its tentative colonial enterprise in Africa (Ethiopia, Eritrea, southern Somalia, and Libya).

A second break in the silence on malaria conditions in Ethiopia came in 1941–1942, with a team from the Mobile Malaria Section of Britain's Army Medical Corps, which visited landscapes of several sites around Ethiopia, reporting "spleen rates," mosquito habitat conditions, and types of mosquitoes the group captured.[9] While the "flying" visit included a tour around the "malarious" Rift Valley, it did not include the Blue Nile Basin. That Rift Valley visit does, however, tell us about an Ethiopia with no formal health system outside of Addis Ababa and a few mission stations (see chapters 2 and 4). No antimalaria work was in place—in those terms Ethiopia remained virgin ground. Malaria had galloped and then receded within its historical patterns, but left only sketchy reports. For some places there were no records at all. In some areas quick visits by specialists relied on researchers poking beneath children's rib cages to try to detect swelling of the spleen, a sign of a previous malaria infection.

Malaria rates as measured by spleen exams (the main malaria indicator used) increased as altitudes declined. The dedicated malariologist Giaquinto Mira, who had stayed in Ethiopia after the Italian occupying forces left, reported a strong correlation between altitude and spleen rates in the eastern highlands (he did not report on the Blue Nile watershed):

Table 3.1. Spleen Rates in the Eastern Highlands (1941)

Place	Altitude	Spleen Rate
Hirna	2,000m	4
Gelemso	1,900m	14
Bedessa	1,800m	42

The low spleen rates show an area's limited exposure to malaria. Higher rates would show malaria's mark. These results show unstable malaria and confirm local knowledge and who lived where—a human geography that understood the altitude effects of malaria, showing malaria as endemic only at eighteen hundred meters and lower.[10] Mosquito collection from a site also added to guesswork about malaria's presence. And, above all, the British army team's eleven-month visit in 1941–1942 concluded that malaria was seasonal and the main vector was *An. gambiae*, which they found "throughout the area," but on seasonal breeding sites like flooded lakes and river margins. Other mosquito types, including "true lake breeders," like *An. pharoensis* and *An. funestus*, were then only "suspects" in the hunt for malaria's causes and culprits.[11] There was still much to learn.

Episode Two: The 1953 Epidemic, Malaria Modern's First Meeting

The year 1953 was a benchmark in Ethiopia's malaria history, a specific malaria event in the Blue Nile watershed and the human response to it. That year saw another big event in Ethiopia's public health history—the opening of the Public Health College and Training Center at the old imperial capital Gondar, at the northern end of the Blue Nile watershed. The college's first classes were held in October 1954, a year after Haile Sellassie's government officially joined the U.S. Point Four program and U.S. military assistance replaced postwar British involvement. The Gondar Public Health College would eventually admit annually 30 health officers, 30 nurses, and 30 health trainees to its programs—a turning point.[12]

But the college was a year too late. None of these trained medical personnel were there in 1953 when a major malaria epidemic hit the Lake Tana area. It was as though a tree had fallen in the forest. Local people felt the effects of fever and death, but internationally the suffering fell on deaf ears, even in Addis Ababa, as a malaria epidemic swept across the northwest region. The world heard nothing of the event, and even local memory recalls little of it. Sir Gordon Corvell, who visited Ethiopia—including the Lake Tana region—two years after the outbreak reported on the severity of the human effect, but also on the sketchy details left in its wake:

> In most years malaria in the plateau surrounding Lake Tana is not a serious problem. The main transmission season which occurs immediately after the monsoon rains, in October and November, is a short one, and consequently very little immunity to the disease is built up among the population. It follows that in years where the conditions are particularly favorable for the production and longevity of the vector species of mosquito, this region is liable to be visited by severe epidemics of malaria attended by a high mortality.[13]

Such an epidemic occurred in 1953, the year following the author's survey of Bahir Dar. He then tells us about that public health disaster:

> In this year it is said that there were 7,000 deaths from malaria in the plain between Gondar and the lake, and that in Kolladuba 300 died out of a population of 1,500. As to the frequency with which these epidemics occur, particular inquiries at Gondar, Kolladuba and Gorgora in 1955 failed to reveal any information as to the date of any previous epidemic in the area.[14]

Apparently, ideas about the "fatal zone" had fallen out of the memory of both malarial science and local recollections.

There are two remarkable bits of malaria history here. First, there were no local health facilities in the wide area of the epidemic, but the estimates of infection and death were of an event on a massive scale. Second, although the World Health Organization began in 1948, it had no mandate to act in colonial Africa, and it had relegated Ethiopia—only briefly an Italian colony—to the edge of its Eastern Mediterranean zone. While key parts of a new era of Ethiopia's international relations emerged in 1953, it was too late. Nature had intervened: A deadly epidemic of malaria had struck the

Blue Nile Basin, though making few ripples that reached Geneva, London, or even Addis Ababa. Out of sight, out of mind.

A third "fact" about the 1953 epidemic was a brief mention in Gordon Covell's 1955 report of his previous visit to the area (the sparsely settled Bahir Dar lakeside shoreline) in 1952. His survey was, in fact, directly a result of a plan to construct a city at the site of the Blue Nile's exit from the southeast corner of the lake. Economic policy folks in Addis Ababa had launched an ambitious scheme to establish a city, which would be called Bahir Dar, smack in the midst of the "fatal zone." In Corvell's pre-epidemic visit, his team had collected 18 species of mosquitoes (7 anopheline and 11 culicine), but he recorded no local memories of malaria outbreaks. Strange.

Though Covell reported horrific figures for Blue Nile Basin malaria deaths in 1953 — 7,000 deaths northeast of the lake and 300 deaths in the Kolladuba village near lakeside (1 in 5 of the population of 1,500), he could find "no reliable information" in those sites when he asked about previous epidemics in the area.[15] A strange testimony on Covell's field research, but perhaps also on the lack of memory within local communities who had suffered symptoms and deaths, but had not attributed them to clinical evidence of a disease diagnosis. Asnakew Kebede, a veteran malaria specialist, who was born in the area just after the 1953 outbreak, recalls that few relatives or neighbors remembered details of it, since it fell into the longer-term pattern of occasional epidemics and little exists in the form of a written record.[16] Were somehow Blue Nile watershed malaria epidemics considered "normal," since they were unremarked upon in memory? We can only guess about the answer.

One local oral account of the epidemic gives an imaginative, human sense of where the epidemic had its origins. This was Asres the cleric/magician, whom we met in chapter 2 and in this chapter's opening epigraph. His remarkably detailed and imaginative memoir tells us:

> In 1946 (1953 A.D.) a fever engulfed the country from Metemma [a town on the Sudan border]. A doctor died in Dembia [north of the lake]. Caregivers were ordered to work in teams of two: if one died the other survived to complete the work. Eleven people died in a single day on the river Djenga Goumara. The people who buried them were so weak that three of them died as well. Others were dug up by animals because the layer of dirt that covered the bodies was thin.
>
> When the disease came to the village of Sendegue, it left corpse upon corpse. The survivors fled. In their haste, they had time to take only a few precious possessions. Everywhere baskets of leather,

hides which served as beds, belts, and bottles were abandoned. The hyenas scraped the bones, walked the grounds and entered the homes. They even pulled the carrion inside to devour them at their leisure. If by chance a calf or a cow came by, they attacked biting its throat. Only the most curious came out during the day. The others stayed inside the houses and made them their homes.[17]

This memory of malaria's effect on humans is clear. Asres's account offers us a sense of the chaos that malaria could visit on a local human ecology. But are these recollections in cultural memory also symbolic, or are they real?

Yet the story also quite accurately charts the malaria outbreak's movement from lowland hot spot into the highlands in a classic pattern seen before and since. People had not expected malaria in the highlands.

The historical record is sketchy, at best, and disturbing in local folks' sense of helplessness. In this image, malaria's outbreak had come from Metemma (a town on the Sudan border lowlands), from which the sickness reached the highlands. For Asres and local folks who followed him, malaria was a force of nature and her spirits. Five years later, epidemic events and those forces of nature played themselves out again.

Episode Three: 1958's Benchmark Bees in a Smoked Hive

> *In retrospect it appears that the malaria epidemic of 1958 began in late June, although no one in the medical services at the time knew or suspected that an epidemic of unusual proportions was in its beginning stage.*
>
> —Russell Fontaine, WHO malariologist (1959)[18]

June was very early for signs of a malaria outbreak, even around the lake. But in spring 1958 there had been unusual rains and higher temperatures with higher relative humidity, ideal for *An. gambiae* mosquito breeding. Average daytime temperatures were higher than ever recorded in the highlands; the same was true for relative humidity. In the dry season relative humidity usually is less than 50 percent, but in the 1958 dry season it was closer to 60 percent or more. Mosquito larvae, especially *An. gambiae*, grew fast in the unusual humid conditions and emerged out of rain-fed puddles as blood-hungry adults. That spring, by sheer luck, a group of U.S.-funded field researchers, doing a routine malaria survey, found themselves in the middle of an unusual June outbreak. And as we later learned, these mosquitoes were

An. arabiensis, which specialized in quick recovery of their population after droughts and dry seasons.

The researchers began to collect blood from victims of fever. In one village 35 thick-film blood examinations showed 22 cases positive for *P. falciparum*, 1 for *P. vivax*, and 3 for mixed infections. In another village, close to Bahir Dar at the lake's edge, 48 tests showed 10 infections. A school and a local prison (both captive audiences) showed 10 positive cases out of 137 examinations. For all three of these sites, the combined rate of slide-confirmed infection was 31 percent! Astonishing. And in June?

The imperial government and its international partners, WHO and ICA (the precursor to USAID), began a belated and desperate campaign to respond by recruiting a motley team of health workers from Blue Nile/Lake Tana area hospitals and public health agencies; there were also missionaries, police, Ethiopian Red Cross workers, and Gondar Public Health College students. The work would be a challenge: There were no paved roads and few vehicles, so the workers traveled by foot and by mule.

The care they were able to offer was hit-and-miss. Sometimes in remote areas the workers gave a single massive dose of 600 mg of chloroquine, and in more accessible areas 300 mg over three days. Eventually they gathered 500 volunteer souls to distribute—willy-nilly—4 million chloroquine tablets to feverish patients in a period of eight weeks.[19] Desperate, scattershot. They treated all fevers as malaria, risking chloroquine drug resistance in the long term.

By the late fall, the still-new Gondar Public Health College had organized small teams of staff and students to visit rural sites and estimate the numbers of the afflicted and the numbers of deaths. It was a grim scene indeed. Russell Fontaine, malaria specialist with the ICA, later estimated overall malaria deaths in Ethiopia for that epidemic year at 150,000.

But the real effect showed itself in the local numbers and human experience. The Lake Tana area's only hospital, a Seventh-Day Adventist hospital at Debre Tabor, northeast of the lake, was at high altitude above the malaria zone, but it received patients who walked or had neighbors and relatives to carry them on stretchers or wooden beds to its "clinic." All told, hospital staff in Debre Tabor consisted of one foreign physician and a nurse. So calling the Debre Tabor facility a hospital was wishful thinking. But those hospital records do us a service by allowing us to compare; they tell us about the number of sufferers of epidemic malaria before the 1958 epidemic and when it broke out. The epidemic's effects could be found in its previous patient records. In the 1948–1957 period, the hospital treated fewer than 126 cases of malaria per year (273 in the 1953 epidemic year).

In three months of 1958, the hospital recorded 2,780 "walk-in" cases of what they diagnosed as malaria, all from districts that the malaria tsunami had washed over; the epidemic had overwhelmed the mostly nonimmune residents.[20]

The most immediate and poignant accounts of the epidemic, however, came from young students and staff of the Gondar Public Health College, who traipsed on foot to distant sites in the lake region and then offered witness passed on in Fontaine's 1961 WHO report. Fontaine offers us only the young men's first names, but their brief field reports conveyed their emotions. They had no treatment to offer, only anguished testimony:

> In late October 1958 Taddesse wrote that: "it is hard to find a healthy person in a family. There are not enough healthy individuals remaining to fetch water for the sick. Many people are dying — as many as twenty persons out of a hundred sick. The crops which are unattended for lack of guards are being destroyed by baboons, wild pigs, birds and other wild life."
>
> On October 13 Mamo sent word that "we treated many malaria patients at Bahir Dar — everywhere we went we met cases of the same complaint. . . . We saw many patients in their bed; practically everybody is sick. Over 500 died since the epidemic outbreak three months ago. The people there said there is a slight epidemic every year, but such a serious outbreak was only known about thirty years ago."[21]

And one more poignant witness:

> Mogues, working in the districts southeast of the lake wrote: "In each house we were able to find three or four patients who complained of subjective symptoms, such as chilling, severe headaches, sweating, pain in the back and extremities. The objective symptoms were jaundice, anemia, emaciation, high fever. After four or five relapses, the headaches, sweating, and pain became unbearable for many patients, who then exhibited a muddling delirium with coma, ending in death. Most of the patients were between the ages of 5 and 20 years; second in frequency were infants and preschool children. Since they are far away from even the simplest clinic, which means no possibility of saving their lives, they are dying like bees in a smoked hive."[22]

Fontaine, the WHO researcher in charge, himself preferred a year later to publish a more dispassionate, ecological explanation:

> Considering the weather factors in the light of their effects on the propagation, longevity, and dispersal of the principal malaria vector mosquito A. *gambiae*, it is evident that all three would be enhanced. Such factors would also favor development of the extrinsic cycle of the parasite P. *falciparum*. Taking combination of the vector and parasite operating under optimum conditions of development in a large, susceptible human population, we have then the essential ingredients for the generation and explosive extension of an epidemic such as was seen in 1958.[23]

Fontaine also recalled Covell's report on his 1929 Sinda, India, experience, noting that mosquito abundance (which he attributed to climate only) and a low-level community immunity (as in highland Ethiopia) promoted the epidemic spread of malaria via the mosquito An. *culifacies* (a primary malaria vector in India, but not in Ethiopia). Fontaine's view saw the epidemic as an example of entomological ecology: Malaria was a product of complex interactions within a natural world that included climate, humans as malaria sufferers, and malaria parasite reservoirs. Humans' dual role was a part of the mystery of malaria.

But there was more: On the report's third page, in the middle of a paragraph, Fontaine tells us that

> the only malaria free district in the large territory was the DDT residual sprayed Dembia plain, an ICA-assisted malaria pilot project covering an area of 2,500 square kilometers on the north end of Lake Tana. Only 80 cases were reported from the project with an estimated population of over 60,000, and there were *no* deaths attributed to malaria. . . . That DDT spray project, which had suffered in the path of the 1953 epidemic noted above, had begun in June of 1957 when all dwellings were residual sprayed with DDT at the rate of 2g of DDT per m^2 of wall space.[24]

Fontaine's almost casual observation, buried in mid-paragraph, was to become the foundation of Ethiopia's next half century of malaria control (see chapter 5).

Episode Four: Discovering a New Dog in the Hunt (1964)

> The author reports on some 200 laboratory crossing of 36 strains
> of Anopheles gambiae from many different parts of Africa which
> shows the existence of five mating-types in what was until recently
> considered a single species.[25]

Indeed a game changer. Professor George Davidson's dour 1964 description of his discovery of "mating types" of *An. gambiae* mosquitoes in his London insectarium produced only a barely audible subterranean tremor in malaria studies at the time, but eventually would transform our understanding of malaria in the Blue Nile watershed, in Ethiopia—and malaria in Africa as a whole. What he originally called "mating types" he concluded two years later were actually distinct species that lived, reproduced, and transmitted malaria quite differently. But Davidson's lab results took almost half a century to have their true effect in policy and practice of the antimalaria efforts on the ground. The movement of that knowledge from the laboratory on Keppel Street in Bloomsbury, London, to policy, then to malaria workers in the Blue Nile watershed, was a long process. It took more than a decade for that tremor to spread. And it took another four decades for science, policy, and antimalarial field activities to appreciate fully the behavioral differences in these "mating types."

Of the five species of anopheline mosquitoes that Davidson isolated, one of them, Species B (later known as *An. arabiensis*), behaved in a way significantly different from the Species A known in West Africa and humid zones. There was a new game in town—a new dog in the hunt—though, ironically, she had perhaps always been there. It is just that human and malarial science had not recognized her. And it was unclear who was hunting who.

The small, elusive flying insect that had plagued Europeans in West Africa and threatened African infants until they acquired immunity—or died—looked to the naked eye like the same mosquito found in Ethiopia and much of arid Africa, but her behavior, survival skills, and life rhythms were quite different. The *An. arabiensis* mosquito (aka Species B), it turns out, had co-evolved with humans and had adapted herself to thrive in the semiarid zones of Africa. She sought animal as well as human blood, and she sometimes took blood meals outdoors (though she rested to nourish her eggs indoors), and she could survive dry season conditions. How she did it is still anyone's guess.[26]

And she had one additional trait: The rebound of her numbers came quickly after the first arrival of rains—almost miraculously so.

Yes, *An. gambiae* Species B would bite at different times than its West African cousin (Species A), have a taste for mammalian and not just human blood, and could virtually disappear during dry seasons. But then she would inexplicably jump back into play in full force only a few weeks later. These behaviors would all be conundra for the antimalaria effort that had for decades unknowingly focused on a quite different mosquito.

In fact the experience of a seasonal ecology of malaria "season" in Ethiopia was very much tied to the adaptations in Davidson's London laboratory of the newly "discovered" *An. arabiensis*. Malaria had not changed with the 1964 publication of Davidson's work, but that 1964 work was a major turning point in humans' understanding of the nature of the malaria vector that has dominated the Blue Nile watershed and Ethiopia as a whole. The mosquito was, we found yet again, not a fixed enemy with rigid defenses like a line of pawns in a chess game, but she was an adaptable foe—a guerrilla fighter—dancing away to a new tune whose survival strategies had exposed her and her sisters more clearly once Davidson turned his attention to his breeding laboratory results back in 1964. But the rub was the question of how quickly the antimalaria bureaucracy on the ground in Ethiopia and the Blue Nile watershed would change its habits and strategies.

The answer was: Not very quickly at all. It took almost four decades for antimalaria policy change to play itself out in the field. Despite the rather dramatic evidence of the new vector, she first appeared in research and policy changes almost a decade after she made her appearance in Davidson's London school laboratory in the early 1960s, when he hatched eggs from different African sites (including Bati in northeast Ethiopia) and got five non-crossbreeding types, one of which he called "Species B." In his earlier 1962 article, he first hesitated to state that this was a new species, but two years later, in 1964, he published his findings in the *Bulletin of the World Health Organization*. In that article's postscript and final paragraph, he boldly told us that the author (i.e., Davidson) "is now of the opinion that all five mating types should be considered as true biological species."[27] Voilà—the new kid on the block.

But the more important question still remains: How long did it take for this research revelation to filter into policy and common knowledge? The December 1965 Quarterly Report of the Ethiopian Ministry of Health noted, rather briefly, that "Species B of the A. gambiae complex" had been identified from the live eggs sent routinely to Davidson's lab from several sites in Africa.[28] To the antimalaria world this fact was an outlier mentioned only in passing and had no apparent effect on antimalaria strategy. A 1970 journal article by two top Ethiopian malaria scientists, "Malaria Survey in North and Northeastern Ethiopia," makes no mention of Davidson's work on Species B, perhaps because the antimalaria world's line of sight was focused on the parasite but not the mosquito—the notoriously elusive delivery vessel.[29] She was of less concern to them, perhaps since they expected DDT spray programs to achieve eradication. Other reports from WHO, USAID, and visitors in that period of the late 1960s and early 1970s make no mention of *An. gambiae* Species B, or its key behavioral differences from its humid zone cousin, Species A, soon to be renamed *An. gambiae s.s. (senso stricto)*.

The very first reference to Species B of the *An. gambiae* "complex" in the bureaucratic literature was in the July 1968–July 1969 report to WHO, in Ethiopia's own Malaria Eradication Program Annual Report. That report's primary message was about the difficulty of getting local cooperation from farmers whose antitax rebellion in the Blue Nile watershed had also caused them to resist worker efforts to collect diagnostic blood smears. Farmers' anger over a new land tax clearly superseded any motivation they may have had for cooperating with the government on antimalaria work. Malaria was hardly their priority. Those folks were highlanders for whom malaria was an afterthought.

Yet that 1969 report also offered two interesting details that foreshadowed a recognition of the *An. arabiensis* (i.e., *An. gambiae* Species B) in Ethiopia. First, the report noted "the very local nature of malaria." Those mosquitoes bred in small, sun-lit, clear puddles that disappeared in the dry season. In other words, malaria's unstable nature in Ethiopia varied from year to year and place to place. More importantly, the report acknowledged that the two most important vectors were the *seasonal* prevalence of "*An. funestus* and *An. gambiae* (Species B)."[30] The report was unintentionally insightful since it recognized that the problem of disruption of antimalaria actions could be political—a tax rebellion—and also noted the *local* complexities of season and the presence of a new species. The dance was indeed complicated and moving in new directions.

Finally, a November 1972 report of the National Malaria Eradication Service (Ethiopia) noted that Species B was in Ethiopia and might have behavioral differences from other members of the *An. gambiae* complex. But the report added a caveat:

> The observations made so far about the behavior of different species of the *An. gambiae* complex are not yet sufficient to determine definitive differences in feeding and resting habits. Generally

speaking, observers believe that *An. Gambiae* A is more anthropo-
philic and endophilic [preferred human blood and indoor biting]
than other members of the complex. *An. gambiae* is regarded as
being catholic in its feeding and resting habits since it has been
observed feeding on men and animals and resting indoors as well as
outdoors at random.[31]

Catholic in its feeding and resting habits? Were mosquitoes really so flexible?
Policies on malaria control needed to catch up with what the mosquitoes
were actually doing. Indeed, her habits and elusive character made it a new
game, or rather a new dance altogether. The Valkyries were, however, now
in plainer sight.

CHAPTER 4

Tragedy of the Jeep, 1958–1991

Hope and the Return to the Drawing Board

ON SEPTEMBER 30, 1964, symbols of hope arrived at the Red Sea port of Assab; to be exact, 108 of them. They were sleek, modern, and forest green. They were American-made. They were Jeep Wagoneer pickup trucks destined for Ethiopia's Malaria Eradication Service; the decal of a shield with American flag colors and a clasped handshake (later over-painted as MES—Malaria Eradication Service) would soon be emblazoned on their doors. A new generation of malaria eradication specialists recruited from the best and the brightest of high school graduates would take the Jeeps to the far corners of the modernizing empire. Hopes were high since these vehicles and their drivers would be the spearhead of the Ethiopian government and the World Health Organization's goal to eradicate malaria by 1980 as a prerequisite to Ethiopia's economic development. The total cost of eradicating malaria, according to the plan, would be US$50 million.[1] The Jeeps would carry a symbolic hope, but also polished aluminum spray canisters, boxes of powdered DDT, and staff trained to follow an elaborate plan to spray houses across the country.

Malaria was well known as a disease of the lowlands, not the highlands, but Ethiopia's development plans called for towns laid out in postmodern grids, and populations of workers to settle lowland areas for producing cotton, sugar, sesame, pepper, and other goods intended for world markets. Those crops came from lowland-loving plants, and lowland-loving malaria awaited those workers and administrators who dared try their luck. It was a cruel contest, since most migrants hoping to earn cash wages had no choice but to risk moving to lowland sites. Employment meant economic survival.

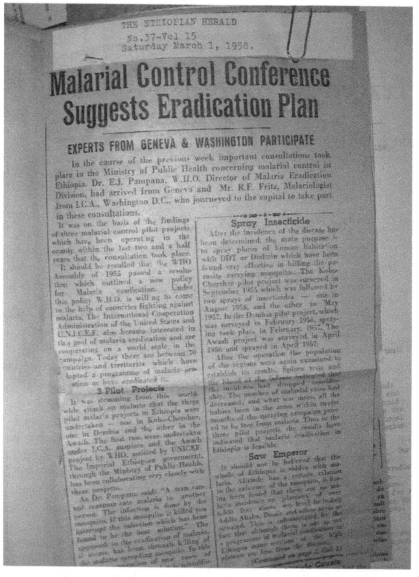

Figure 4.1. Eradication optimism. (*Source: Ethiopian Herald* [1958]. Photo by author.)

Those forest green Jeeps became a generational marker of hope, of the determination of applied science, and, ultimately, of a qualified failure. Malaria would live to dance another day.

Figure 4.2. Transporting the eradication dream. (*Source*: Ethiopian Ministry of Health. Photo by author.)

The 1964 arrival of those 108 Jeeps, was, in fact, a point of high optimism in the human-malaria struggle in Ethiopia. The regiment of antimalaria vehicles lined up for shipment to the Ministry of Health in Addis Ababa also signaled the training of a cadre of young elite high school graduates who were the shock troops of a campaign that had begun in 1959 with the formal announcement of Imperial Order number 22 that established Ethiopia's Malaria Eradication Service (the MES insignia that later emblazoned the new Jeeps in 1964); an earlier insignia was of a clasped hand with an American red, white, and blue shield, but even the over-painted U.S. flag signs had worn off. With that 1958 conference and 1959 law also came the opening of a training facility at Nazaret (a site ninety-eight kilometers south of the capital) that began to prepare these optimistic young men (yes, they were all men) who would venture forth to rural places to fight a disease that threatened to confine Ethiopia's economic growth to malaria-free highland zones above two thousand meters in altitude. Ethiopia's economic planners knew all too well that new farms, plantations, and settlements were in dangerous lowland areas that would have to be malaria-free. And few, if any, foresaw the slow creep of malaria into higher elevations—new lands for the malaria chess match.

The team to engage the enemy and the unfolding chronology of events reflected a growing engagement of public health programs in a post–World War II, Cold War world. It included the World Health Organization, the

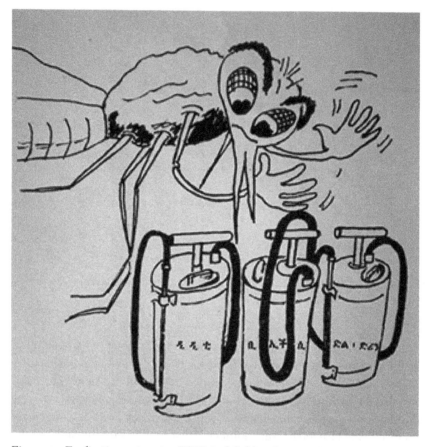

Figure 4.3. Eradication, mosquito, DDT, and dieldrin from a 1964 Ministry of Health pamphlet. (*Source*: Ethiopian Ministry of Health. Photo by author.)

International Cooperation Agency (in 1966 renamed USAID), Ethiopia's Ministry of Health, and its shock troops, the MES. But it also included the bodies of the migrant workers, families, and others who braved Ethiopia's malarial zones that offered high economic potential in the forms of employment and financial rewards.

Eradication: 1958–1969

The years 1958 through 1969, more than a full decade, had begun with the horrendous shock of the 1958 malaria outbreak of stunning proportions: In the empire as a whole, in the 1958 epidemic year an estimated 3,000,000 came down with the disease and 150,000 died of it.[2] In the Blue Nile Basin around Lake Tana (where the young Mogues observed the 1958 tragedy on foot and without medication to offer), helplessness was all too obvious locally, but invisible to the world.

With the 1958 epidemic, malaria's power to kill in Ethiopia's ecology was all too apparent to its government and to international partners like the World Health Organization and the U.S. International Cooperation Agency. And it was all too apparent to those who suffered from periodic, but unpredictable, epidemic outbreaks and some seasonal cases. Yet, from the first 1959 announcement of the antimalaria program, it adopted eradication as its mantra and mosquito control through household spraying as its method.[3]

At the outset, choosing to spray was the obvious choice for holding back malaria and then eradicating it. The MES and their WHO allies were optimists. The Ministry of Health itself, however, had little to offer: no local health stations or clinics, or even roads to reach malaria outbreak areas or the houses that needed spraying there. But it had spray equipment, the chemicals, and energetic MES workers who walked carrying bright silver canisters full of DDT (mixed for 2 grams per square meter) and another chemical, dieldrin, to spread out across a vast landscape of vulnerable houses.

And they had learned from other world areas that the cheapest way to control malaria was to spray houses to kill or at least suppress the dreaded disease vector, the anopheline mosquito. There were two poisons of choice: dieldrin (a chlorinated hydrocarbon banned in 1974); and dichlorodiphenyltrichloroethane, aka DDT. Both chemical compounds worked for positive results in *kdr* (knockdown rate) tests, but DDT was far and away the longest lasting—and the cheapest. Wikipedia gives the shorthand version of how the chemical worked:

> DDT opens sodium ion channels in neurons, causing them to fire spontaneously, which leads to spasms and eventual death [for the mosquitoes].[4]

From 1958 through the first decade of the twenty-first century, DDT remained the dominant spray ingredient used in Ethiopia. In the United States and in much of the world, Rachel Carson's observations of DDT's effects on the beaches and woodlands of New England brought cautionary alarms about DDT's effects on the food chain. In Ethiopia, however, its effectiveness outweighed its potential damage, and WHO, USAID, and Ministry of Health programs kept it in play well into the twenty-first century. Fieldworkers could mix DDT powder with water, fill their backpack sprayers, and walk to the targeted houses, towns, and villages. By systematically

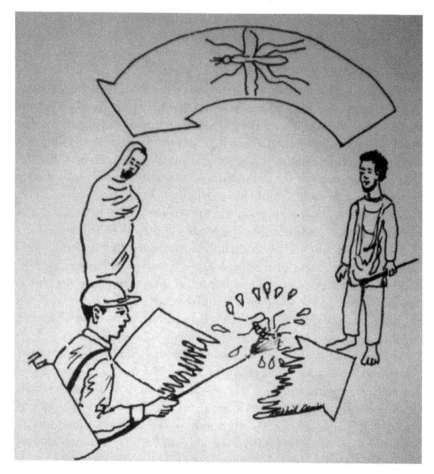

Figure 4.4. Spray campaign of 1964: Breaking the cycle. (*Source*: Ethiopian Ministry of Health. Photo by author.)

spraying the house walls with a well-rehearsed set of dancelike steps, moving their spraying wands at prescribed angles up and down, shuffling their feet forward and back, left and right, they could cover a wall quickly and efficiently and move on to the next thatched house.

Results were quite encouraging. In 1959, Ethiopian malaria entomologists performed a knockdown study of the effectiveness of DDT on mosquitoes resting on an interior house wall—a simulation of mosquito behavior after taking a human blood meal. In this test, 99 percent of the mosquitoes died. A triumph for science: science 99, mosquitoes 1.[5]

In other studies that compared sprayed areas with areas that felt the full effects of the 1958 epidemic around the Lake Tana basin in the northeast, it seemed that a battle had been won. In July through December 1958, in four provinces with a population of 8,000,000, malaria had struck with morbidity rates exceeding 75 percent in places, with total deaths over 100,000. And the villain was *P. falciparum*, the most deadly parasite to appear that year in almost three-quarters of all cases (though only 57 percent in other years).[6] But in the Dembia plain northeast of the Lake Tana area where a pilot project had sprayed with DDT, there were only 80 cases among a population of 60,000.[7] Remarkable. In using indoor residual DDT spray, malaria eradication for Ethiopia had its first magic bullet. Or so it seemed. Unlike Facist programs in Italy that had the means to distribute quinine widely to peasant farmers and workers, Ethiopia had no such ability to deliver their DDT bullet.

What these data seemed to show, and what optimists drew upon, was the effect of the DDT spraying through the remarkable decline in the number of malaria cases between September 1956 and September 1959. This was particularly true when, during the 1958 epidemic, the sprayed zone remained virtually malaria-free.

TABLE 4.1

Human Malaria Indices before Residual Spraying and after Dembia Pilot Project

Date of Survey	No. Blood Exams	No. Positive	% Positive	P. *falciparim*	P. *vivax*	P. *malariae*	Multiple	Unidentified
October 1956	749	93	12.3	68%	14%	6%	2%	10%
June 1957	293	9	3.0	67%	11%	11%	-	11%
November 1957	824	5	0.61	80%	20%	-	-	-
August 1958	69	0	0.0	-	-	-	-	-
September 1959	909	21	0.44	67%	-	-	33%	-

Three DDT spraying cycles were applied: May 20–June23,1957; May 1–30, 1958; May 1–June 7, 1959

Source: Abdullah E. Najjar and Russell E. Fontaine, Dembia Pilot Project, Beghemder Province, Ethiopia, Second Regional Conference on Malaria Eradication, Addis Ababa, November 16–21, 1959. WHO/MAL/265/EMRO/MAL/41, WHO Reference Library, Geneva.

Amazing results, since the study overlapped almost exactly with a deadly epidemic whose effects in the wider zone we have already seen. Yet the good news of this pioneering study in the Dembiya pilot region had to be scaled up—an enormous challenge to science and to bureaucracy. In the chess game there seemed to be an opening for an attack in response, but malaria moved in many directions, in a dance with many complex, overlapping twists.

Since malaria appears with a wide variety of the anopheline mosquito types around the world, fieldworkers since the mid-1930s, Italians during the Fascist period, and WHO workers in the 1950s had already done remarkable work identifying the types of mosquitoes in the Ethiopian region, as they had done for Italy itself in the first third of the century. A British survey in the 1940s had added to that work, so by 1958 it was pretty widely known in mosquito circles that Ethiopia had thirty-one types of anophelines in quite specific Ethiopian ecologies and with quite different abilities to carry malaria.[8] Some, like *An. funestus*, liked permanent water sites such as marshy lake shores; others, like *An. pharoensis*, bred well in slow-moving rivers (see chapter 7). *An. arabiensis* was ubiquitous—she appears in almost all malarious sites. Italy had had only one important species.[9]

By 1958, there was a rather clear consensus that *Anophelene gambiae* was for Ethiopia, and most of Africa, the ubiquitous and most dangerous vector for malaria. Most importantly for the antimalaria program, that 1959 series of lab tests had shown that the DDT knockdown rate was close to 99 percent (i.e., the number of DDT-sprayed mosquitoes killed). For the alternative but more expensive insecticide dieldrin, the results were as good or better, but both were above 99 percent effective.

DDT was indeed the magic bullet. Or at least it was the only affordable one, and therefore the chemical of choice for malaria eradication. MES and their partners in ICA and the WHO moved into the 1960s with confidence, knowing that they had in their holster a cheap, effective, and long-lasting weapon to combat their enemy. Potential checkmate, they thought. The MES also had a GR (geographic reconnaissance) scheme that offered the broad brushstrokes of a national map divided into zones, which they set in priority as A, B, C, and D (see fig. 4.5), a template for planning their pre-eradication strategy. Their goal was a set of stages leading to eradication.

In effect, this map was a set of plans for malaria eradication, but it was also a map of plans for a modern political ecology and economic potential for the modern nation. The MES and its allies chose to concentrate its resources on Area A, an area of the highlands and part of its periphery that was, in effect, historical Abyssinia (i.e., the Christian highlands and areas they

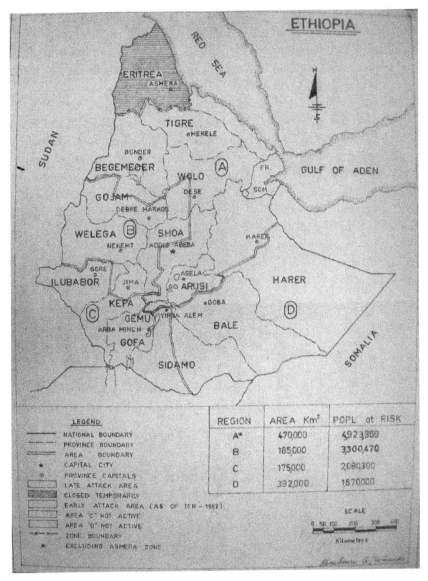

Figure 4.5. Regional map for spray program (1967). (*Source*: WHO Archive, Geneva. Photo by author.)

dominated in the Haile Sellassie era). But it also included strategic areas for economic development in the Awash Valley (cotton, sugar, citrus); north-west borderlands (sorghum and sesame oil and seeds); and the Rift Valley (red pepper, sugar, haricot beans, soybeans, commercials fruits) that made up much of Ethiopia's potential global agricultural exports.

Ethiopia's malarial eradication map laid out its part of their (naive) global chess game—move by move. The moves were (1) preparation; (2) attack; (3) consolidation; and (4) maintenance, leading to eradication (checkmate!). For Ethiopia, fighting malaria was the primary national program of the Imperial Ministry of Health, and its importance meant it had precedence over health goals that might have included rural health centers, maternal health, other infectious diseases like polio and measles, and education. Malaria eradication was the primary goal because they believed they could achieve eradication and then move on.

These maps, and others, were parts of what WHO and its Ethiopian allies called the attack phase of malaria eradication. The other elements were:

- Geographic planning to set priority areas for spraying
- Drugs (chloroquine, primaquine, and quinine) for treatment (not prevention)
- DDT powder (for mixing with water on-site)
- Spray equipment (backpack canisters and DDT mixing tools)
- Manpower (trained field staff and managers from the program at Nazreth)
- Mobility (i.e., the green Jeeps of optimism)

There were, however, some signs for concern even then. Dr. Joliet, the WHO entomologist who presented the long tables of results for each area and each type of mosquito and each concentration of the poison, did so in excruciating detail. Yet he had results only in cases that had never been sprayed or that had been sprayed only three to four times. He concluded:

> The three species tested to insecticides (A. *gambiae*, A. *funestus*, A. *pharoensis*) are extremely susceptible to DDT and Dieldrin in the unsprayed areas and normally susceptible in the sprayed ones even after 3–4 years of spraying.[10]

What he had not anticipated was that spraying in Ethiopia would continue for more than five decades. The mosquito's dance had, sure enough, included adaptation to chemical spray.

In the United States global malaria was a target of visions for international development. In 1962, plans were underway to fulfill President John Kennedy's interest in malaria that had begun in the late 1950s while he was a junior senator. By 1962, John Glenn and his Mercury mission had orbited the earth three times and WHO had undertaken three rounds of DDT spraying in targeted areas of Ethiopia. MES and WHO had taken moves in their chess game, but unfortunately malaria and the anopheline mosquitoes were playing it as a dance, an ecological engagement in constant motion.

Kennedy's two 1962 projections about science (malaria eradication and a moon landing) were somewhat prophetic by the end of the 1960s decade, though he did not live to see either. In 1969, one happened, the other didn't. When Apollo 11 made its July moon landing on Tranquility Base that year — a triumph of technology over natural barriers — the World Health Organization announced, with chagrin and honesty, that its goal of eradication of malaria was unattainable.

These two 1969 events in the world of science were truly remarkable, but apparent to few, if anyone, at the time. Technology's reach in 1969 had seemed unlimited, except where solutions were more complex than shooting a projectile into empty space (and returning it). Malaria was infinitely more complex.

These were the elements that made up the talented and committed team of the MES program that in 1964 gladly received the 108 Jeeps (and some half-ton trucks), and on the other side were clever mosquitoes and elusive parasites. Mobility for the "Attack Phase," as the WHO described it, was the strategy.

Eradication was the heartfelt goal. The methods were modern, part of a global effort that merged with programs like WHO in what now had become postcolonial Africa. For Africa (and Ethiopia), unlike other world areas, pre-eradication measures came first.[11]

Drugs: Malaria in the Age of Chloroquine

> Sulfonamides can cause numerous adverse effects. Agranulocytosis; aplastic anemia; hypersensitivity reactions like rashes, fixed drug eruptions, erythema multiforme of the Steven Johnson type, exfoliative dermatitis, serum sickness; liver dysfunction; anorexia, vomiting and acute hemolytic anemia can also occur.[12]

Even though these potential side effects of the malaria drug known by its trade name Fansidar (pyrimethamine and sulfadoxine) sound unattractive,

Figure 4.6. A 1964 optimist's view of eradication: DDT spray. (*Source*: Ethiopian Ministry of Health. Photo by author.)

they were among the prices to pay for treatment of malaria when it struck—or seemed to have struck.

Ethiopia's Malaria Eradication Service worked closely with a polyglot stew of malaria specialists from around the world who chased the disease wherever they could find it. The drugs for treating those afflicted with the fever had evolved, sort of, since early travelers had used "the bark"—a boiled infusion of quinine—to treat malarial symptoms, even if they did not know the cause or why the cure seemed to work. As late as 1938, an Italian physician named Aldo Castellani, working in the Lake Tana area, wrote in French to the office of International d'Hygiene Publique in Geneva about his prescription for use of quinine as prophylaxis (i.e., to prevent infection):

> Prophylaxis àvec de la quinine dès les premiers jours de juin jusqu à la fin Octobre en les zones d'altitude à 1.600–1.700 metres. Pour ne pas prendre des quantité inutile de quinine. La prophylaxis pourra êntre en faisant prendre 60 cg. faites trois jours consécutifs de la semaine. Par cette mèthode on obtient des résultes identiques et avec de moindre des quantités de la moindre dérangements que par l'administration quotidienne de la meme quantité de quinine. . . . Prophylaxie chemiquie suivant les mêmes modalities a partir del la deux moitié du mois d'auot et jusqu au mois d'octobre compris, dan les zones qui se trouvent a environ 1.800 metres.[13]

Castellani's surprising recommendation to *reduce* the frequency of quinine dosages showed that the international community was the audience for anti-malaria therapy intended for a French-reading audience in Geneva and the world. It likely came from his field experience with treating Italian officials in occupied Gondar (and in Italy in Mussolini's antimalaria campaigns). Yet Castellani's assertions about malaria's elevation limitations would prove themselves wrong in terms of human patterns and physical contexts of the disease in the changing ecologies of the coming epidemics.

Quinine, or its artificial substitutes, could be used for treatment or even for prophylaxis (prevention). In 1938, the same year of Castellani's Gondar prescriptions, French lab scientists introduced a new drug, Atabrine, which was, in effect, a synthetic quinine. Early on in World War II's Pacific theater, Japan controlled the supply of quinine in Southeast Asia, and Atabrine gained pride of place for Pacific malaria control. It was the drug of choice for U.S. military health workers. Actually, they gave it not to the U.S. Marines and Army, but to the local island populations, who they feared might spread the disease to the occupying military. In James Michener's collection of Pacific War short stories, one memorable character was Atabrine Benny, a medical staffer who was locally famous among islanders for distributing the drugs, which actually turned its patients' skin bright yellow, its most physically apparent side effect. There were other effects, too.[14]

By 1960, the global war on malaria included a number of drug therapies, but the cutting-edge drug that emerged and was most widely used was another synthetic quinine named chloroquine with the trade name Aralen. American Peace Corps volunteers who began arriving in Ethiopia that benchmark first year of 1962 in rural areas recall the pink tablets, pink because the sugary coating disguised the bitter taste of the raw drug. Some also recall its effects on the skin, which for some was the annoying impossibility of cosmetic tanning but for others more serious skin irritation. These Peace Corps volunteers, like me in 1973, and other foreigners, took once-a-week dosages of that chloroquine as a malaria prophylaxis, which either prevented the disease or reduced the symptoms to manageable levels. Those who contracted malaria usually took an increased dosage to suppress malaria symptoms or faced the Fansidar (pyrimethamine and sulfadoxine) treatment, which had potentially dangerous side effects—like exfoliative dermatitis. But even that was better than the fevers and life-threatening falciparum malaria.

For Ethiopians in epidemic vulnerable areas, a lifetime regimen of weekly chloroquine was not possible in practice because of the cost and the drug's long-term side effects. Though foreigners could manage it financially and had access to the drugs, the overall effect on malaria and public health would have been minimal. Through the 1960s and 1970s, however, chloroquine was the drug of choice for the MES, and it was used in response to fever symptoms in an outbreak rather than as an overall panacea to achieve eradication by chemical means.[15]

But a network of pharmacies and health centers outside of towns was still a pipe dream. Diagnostic labs outside of the biggest provincial centers were an even more elusive goal. Clinical diagnosis by health workers or a local pharmacist (i.e., a local shopkeeper who sold drugs with some post–high school training) was grounded purely in intuition and practical experience. Malaria's symptoms were still poorly understood by local folks. Local names for the symptoms were usually translated in clinics or field visits as the disease, malaria: *enqetqet* (shivering); *nidad* (hot or on fire); or just *woba*, which meant "malaria" in several local languages, but could as easily turn out to be typhus (caused by ticks, lice-borne relapsing fever [LBRF]). Even in the 1970s, in nonurban areas there was rarely laboratory equipment and staff to do a blood smear slide that showed parasites in the bloodstream and confirmeda diagnosis of malaria.[16] Given the virtually nonexistent health infrastructure, folk diagnoses in the 1960s were about as accurate as in the nineteenth century—though a seemingly effective drug therapy often actually confirmed an intuitive diagnosis by a local druggist. Using quinine or an extended dose of chloroquine was often effective, but was prescribed mainly by educated guesswork.

But slowly, slowly, aid programs in health education began to change the game, moving some pieces forward. In 1959, the public health college at the town of Gondar (Ethiopia's old imperial capital and what had been an Italian colonial nexus in the 1936–1941 period for the north) had begun to train pharmacists and local druggists whose courses at the Gondar Public Health College, and other similar training centers at the towns of Jimma and Ambo, gave young graduates the skills to open pharmacies (which served also as clinics) in small provincial towns.[17] Classmates from these public health colleges looked across the rural landscape and chose market towns that drew rural visitors on market days and looked like promising sites for pharmacies. These started up with a few drugs or potions to address symptoms of common diseases or to offer ointments for eye infections. The market towns themselves still often had no piped water, electricity, paved roads (or any government-sponsored health care at all). These young medical entrepreneurs opened their rudimentary pharmacies that could dispense antibiotics or even pull an infected tooth in the front of their shop. But those private drug shops were a small step forward.

Bekele Abebe graduated from Ambo Medical Centre in 1968 and looked up the road north and across the Blue Nile. He asked himself which of the emerging road towns had no pharmacy. For townfolk and farmers within walking distance it was a giant leap for health care; for Bekele it was an economic opportunity. His classmates were all doing it, hopscotching into road towns with attractive weekly markets to judge local needs and entrepreneurial possibilities. Bekele chose to hang his shingle in an old caravan market town (Burie in Gojjam) on a picturesque rise 2 kilometers off the main graveled road, 414 kilometers north of Addis Ababa. Near that spot in 1941 the emperor had encamped with his army as he and his British allies pushed the retreating Italians farther south to the capital of Addis Ababa. But Burie in 1968 still had no piped water (it came from the river by donkey), and the town's only lightbulb was in a hotel a couple of kilometers away on the road to Bahir Dar. In town, nighttime light came from candles or an occasional Coleman lantern, or once a month during full moon. His nearest medical competition was a druggist 80 kilometers down the main gravel road; there was a German mission hospital over 180 kilometers north at the shore of Lake Tana. The first two kilometers from Burie was on foot to the gravel road that led to Bahir Dar. Bekele opened his Burie shop in 1968, and in short order his one-room arrangement with a countertop became the five-thousand-person town's pharmacy, outpatient clinic, dentist, post office, and malaria diagnostic center (no lab). On market days he was overwhelmed with rural patients seeking treatment for various ailments.

Malaria was always a possible diagnosis, but Bekele relied on the town's 2,100-meter elevation, the local ecology's constellation of possible symptoms, his practiced eye, and equally on his canny ability to assess where the patient had been, for how long, and in which season. For a place like Burie during most of the 1960s and 1970s, the nearest laboratory or hospital bed was likely more than 180 kilometers distant and a rocky three or four hours' ride away by the small twenty-five-seat Leoncino buses that passed by a couple of times a day. Other passengers would likely include goats, sheep (kept in the cargo bin or on the roof), or chickens. In 1970 those small buses were an important ingredient of the health-care system. In Burie, the sole agent of the government's district antimalaria program was most conspicuous when he appeared in town astride a rather handsome mule.

The MES was not in the business of medical care, and could not be, given the scale of its antimalaria mission and the dearth of other resources. While antimalarial drugs appeared a part of its agenda in the 1960s, the MES could not address health care overall. Its primary goal was to spread out its forces and spray, spray, spray. Silver spray canisters, boxes of DDT powder, and the newly trained staff had hit the ground running—a hit-and-run operation. Was it successful? Would those best hopes—silver spray cylinders and the people in those 108 Jeep Wagons—eradicate malaria?

Despite the obstacles, early results from the spray campaign had been really promising. The international staff that worked with the new generation of Ethiopian malaria workers had committed themselves as a piece of the chess match—an international effort for eradication.[18] The ethnography of the names that appear on letters, reports, and equipment lists suggests a new ethnographic global mix: Najjar, Fontaine, Mehra, Morin, Sonti. An Ethiopian name—Awash Tekle Haimanot—does not appear until 1991 (although talented, low-level Ethiopian field staff were certainly active well before that). Letters and cables came to Addis Ababa's WHO and the ICA office in the U.S. Embassy from ports of call like Cairo, Beirut, Geneva, Brazzaville, and Kampala. Their points of origin overlapped with a period of unprecedented optimism and economic growth in an era of postcolonial independence. If former colonies could burst at the seams with growing cities, trade, and new generations, surely malaria could be a thing of the past as well. After all, it had disappeared (the term then used was "eliminated") in Britain, Italy, and America, and was on the verge of doing so in Brazil, Mexico, and some Pacific Islands.

1969: An End of Eradication?

On July 20, 1969, Tranquility Base called Houston and declared that "the Eagle has landed." I was a high school kid at a Chicago White Sox game, and Walter Williams had just singled over second base. The scoreboard

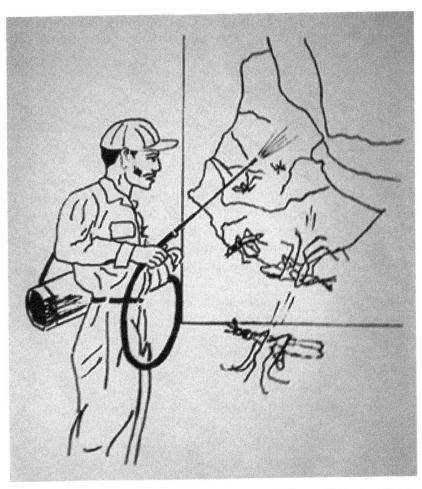

Figure 4.7. Breaking the cycle: Malaria education, 1964. (*Source*: Ethiopian Ministry of Health. Photo by author.)

fireworks went off. Why? There had been no one on base in a meaningless game. Yet a truly remarkable human achievement of science and engineering had just taken place, and had been the reason behind the fireworks sent off into a Sunday afternoon sky. It was a real moment in human history, as Neil Armstrong said when he finally set foot on the moon's surface a few hours later on July 21. One giant leap for mankind. But one can also see it as a rather narrowly proscribed achievement. The baseball announcer in Chicago related the news that Apollo 11 had achieved something amazing. But how amazing was it? After all, the NASA achievement merely involved shooting a rocket at a predictably moving target, landing on a field of inert dust, shooting itself back, and parachuting three guys into a vast body of water. The ecology of the NASA target was simple and static, as was the point of return. Inert dust and an ocean of water. A feat of human talent and engineering, to be sure. And the earth celebrated.

In President Kennedy's second promise in 1962, he had challenged American medical science technology to go to the moon, which had proved much more elusive. Malaria eradication, a global institutional effort, involved a shape-shifting, ecological target that endangered millions, and the human efforts to stem its tide had been going on for about a century with little effect in Africa and in Ethiopia in particular. Really, it had been nearly a century since the revelation of the beast in the body—a protozoan—in 1877, and a few years later discovery of the delivery mechanism, the female mosquito, just after the turn of the century. Science had discovered the biological cause of malaria, though sufferers around the world had known the effect for many years. Now what? A century of laboratory research followed, and the scientific and medical communities adopted an attitude of hope that came from successes over typhoid and diphtheria (both bacteria-caused), polio, smallpox, and even the cattle disease rinderpest (all viral afflictions). Malaria had long surpassed all of these in its human effects, and malaria, by contrast, was a protozoan infection. Surely, it was the next target. John Kennedy and Louis Pasteur had said so.

The best-laid plans . . . In July 1969, at the World Health Assembly, the director-general of the World Health Organization confronted the evidence of the record of its progress toward the goal of eradicating malaria. In May 1970, an SRT (Strategy Review Team) in Ethiopia acknowledged the 1969 WHO declaration of failure and issued a report that faced the reality of the past decade—indeed, the past century—of eradication efforts on the ground. It acknowledged defeat:

> In view of the setbacks and slow progress reported in many malaria eradication programs, the global strategy of malaria eradication was re-examined in 1968/1969 as requested by the Twenty-First World Health Assembly (July 1968). The Assembly expressed its view that future strategy will seek to determine the course of action aimed at eradication best suited to the specific requirements of a variety of country situations. The Assembly reaffirmed global eradication of malaria as the long-term goal and recognized malaria control as a valid and indispensible interim step where *eradication is at present impracticable*.[19]

A global failure. Where did Ethiopia sit in this monumental decision? In a cabled reply dated April 26, 1970, His Excellency the Minister of Health informed the regional director that the Imperial Ethiopian Government agreed to the proposal (in other words, eradication was impractical).[20]

Translated from the deliberately vague global diplomatic language of the 1960s—doublespeak, as Orwell might have called it—a remarkable set of events had happened. But remarkable not in a good way. First, the World Health Organization had, rather subtly, mind you, abandoned the goal of eradicating malaria at the global level. The assembly recommended that governments of the countries with programs underway revise them in cooperation with the WHO and other assisting agencies with a view to adapting them to a strategy calculated to give the optimum results. Second, it had punted the ball to local programs to (to paraphrase) "do the best you can."[21] And for its part, Ethiopia had officially agreed to the strategic retreat, though the MES kept its optimistic door decals, at least until they abandoned the Jeeps altogether in 1980 after almost twenty years of service. Eradication was, in the world of the WHO, a dream but not a practicable goal. The operative word now was *control*. If Ethiopia's snazzy and tough Jeeps did not change their MES decal, their drivers and technicians who rode in them had to accept a kind of defeat and had no choice but to soldier on.

Neil Armstrong had stepped onto the moon July 21, 1969, an achievement for science promised in 1962, but malaria had not left this earth, as had been the promise in that same year—and certainly had not left Ethiopia. *Control* of malaria had now become the much more modest goal for WHO and for practitioners in that country.

In mid-December 1969, the WHO regional public health administrator visited Ethiopia and delivered an ominous message: After a decade of struggle, DDT, drugs, and surveillance,

> it can be said without a doubt that eight rounds of DDT spraying have been unable to interrupt transmission of malaria. There are many factors which are responsible for this failure, some of which are administrative and operational, but others of a technical nature which have not as yet been circumvented.[22]

The optimism of a decade before may have been justified, but the reality of May 1970 gave it a splash of cold water. At that date there was progress to report on the human side: a trained staff of 7,926 malaria workers with spray equipment, the 1964 Jeeps and animal keepers (donkeys, mules, and horses still were important for moving into rural places). The pieces were on the board, but malaria was still in motion.

The record showed, however, that the program phasing that had been laid out so convincingly in the planning documents for the previous five years was barely off the ground. What had been a sequenced and geographically planned frontal assault on malaria had barely moved. The chess pieces were still in place on their original squares, with only a few pawns venturing forward. Moreover, to move forward a bit into the next seven years (i.e., 1971–1978), the political and economic climate would upend the chessboard with an oil crisis in 1973, Ethiopia's own socialist revolution and civil war beginning in 1974, the 1972–1974 famine in the northeast, and the collapse of U.S. aid. Disturbance. These climate factors and political winds suspended USAID programs in health, rural development, military initiatives, one hundred or so Peace Corps volunteers, and related projects supported by European and multilateral donors throughout infrastructures geared toward Ethiopia's planning and development. And there was war with Somalia (1977–1978), civil war with Eritrea, and a northern insurgency that eventually toppled the military government in 1991. These were all catastrophic and unimagined from the point of view of 1970. Malaria loves such disturbances.

In May 1970, half a year after the WHO's surrender to malaria, another international Strategy Review Team spent three weeks reviewing the evidence of Ethiopia's previous decade's antimalaria struggle. The team was a formidable assembly of heavy hitters, and their affiliations were an alphabet soup of stakeholders: IEG Central Laboratories and Research Institute (Ethiopian Imperial Government), USAID, WHO/EMRO, CDC, WHO/Geneva. Their report is a useful benchmark of what was on the ground: past successes and a proposed plan to pivot diplomatically from eradication to control, and from an independent malaria unit to a general Public Health Service where malaria became something of a sideshow—an emphasis but not a priority overall.

This was a major turning point in Ethiopia's public health and malaria strategy. The report laid out three key points that formed the groundwork for a new approach:

1. The general structure of Public Health Services and relation to the Malaria Eradication Service
2. Communicable disease priorities

3. Integration of Communicable Disease Health Services with
the Malaria Eradication Service (MES)

In its own bureaucratic language and inference, this was a 1939 Munich
agreement in which one side of the antimalaria bureaucracy acknowledged
malaria's upper hand in the chess game, outplayed by a nimble opponent
(malaria itself) who played by its own rules.

Part A of the ministry's plan laid out a strategy that, in effect, acknowledged the absence of public health facilities outside of urban areas and
outside of the private pharmacist informal networks that had popped up as
local evidence of private medical practice well ahead of the government's
capacity. It acknowledged that there were 68 health centers and 500 health
stations in existence. But

> in actual practice even those existing health centers and stations
> have been severely limited in the preventative services they can pro-
> vide. Many are staffed by only one team. Shortages in funds, drugs,
> facilities, transportation, and supervision . . . Furthermore, the exist-
> ing personnel are simply overwhelmed by the incessant demand for
> curative services.[23]

The report, in passing, also allowed that malaria operations were limited to
areas below two thousand meters elevation, covering about half of Ethiopia's population and, it may be noted, the most fertile regions appropriate
for major agricultural development (i.e., Area A of the malaria map, see
fig. 4.5).

Planning, at some point, has to meet the complex realities of economics,
politics, and nature's ecology. The responses of nature may have been the
most unfathomable of all. In May 1970, Ethiopia's political revolution still
lay on the horizon, but it was already clear that the plan laid out in the optimism of the early 1960s had gone astray. Here were the players in the game
and their levels of success:

TABLE 4.2
Malaria Eradication Service Employees (May 1970)

	Total	Permanent	Project	Temporary
Headquarters	300	300	—	—
Zone Bases	438	409	29	—
Sectors	7,164	1,176	968*	5,020**
Malaria Training Centers	24	24	—	—
	7,926	1,909	997	5,020

*Includes single-purpose spray squad chiefs and surveillance personnel.

**Includes 2,394 spraymen, 1,741 porters, 276 camp guards, 590 couriers, and 119 animal keepers.

Now let's look at the malaria eradication program's plans for the 1965–1978 period—and its limited achievements.

The political ecology of antimalaria planning, even before the decades of the 1970s and 1980s, shows that the game had barely begun. But let's look a bit closer to the ground at malaria in the age of scientific socialism.

Malaria Meets the Science of Socialism

In 1974, an event took place in Addis Ababa that few thought of in terms of malaria. Disease, and malaria in particular, loves disturbance in a way that archaeologists hate it. For archaeologists, disturbance often erases the past. For malaria, it creates new opportunities for mosquito breeding sites; it disrupts medical care, derails spraying programs, and forestalls public health education, and it also moves in new people as migrants or refugees whose bodies had never before encountered the parasite.

For Ethiopia, the disturbance was political, social, and ecological—a genuine revolution that turned the nation on its head politically and allowed nature and its malarial minions a free run. The symbolic event might have been on September 24, 1974, when, across the road from the World Health Organization offices on Menilek Avenue, a metallic blue Volkswagen beetle entered the front gate of the Jubilee Palace and parked at the impressive front door. Soldiers of the new revolutionary army emerged and escorted a frail eighty-two-year-old man into that modest car and drove Emperor Haile

TABLE 4-3
Program Phasing by Area

as Envisioned in Plan of Operations

Phase	1965	1966	1967	1968	1969	1970	1971	1972	1973	1974	1975	1976	1977	1978
Preparatory	A	B	B	C	C	D	D	—	—	—	—	—	—	—
Attack	—	A	A	AB	AB	BC	BC	CD	CD	D	D	—	—	—
Consolidation	—					A	A	AB	B	BC	C	CD	D	D
Maintenance	—								A	A	AB	AB	ABC	ABC

Actual Situation [May 1970]

Phase	1965	1966	1967	1969	1970
Preparatory	A	—	B	B	B
Attack	—	—	A	A	A
Consolidation	—	—	—	—	—
Maintenance	—	—	—	—	—

Source: Report of a Strategy Review Team, Ethiopia, May 6–27, 1970. WHO7. 0013; WHO 1970–1971. WHO Archives, Geneva.

Sellassie, now deposed, to house arrest in an officer's apartment in a military barracks about two kilometers across the railroad tracks and down the hill toward Nefas Silk, the city's industrial periphery.

Then a political revolution proceeded, violent at times, to unfold in the towns, cities, and countryside. Military officers rose to power briefly, only to fall in quick musical-chair succession to more junior officers. Schools closed when students went on strike, as did taxi drivers, both signs of political and social unrest. At first, the educated class had rejoiced at the prospect of modernization—"scientific socialism" was the term. *Etiopiya Tiqdem* (Ethiopia First). Land reform, and then war in Eritrea, and then the imposition of a military authority with a socialist ideology landed on the bureaucracy, a change that brought many of its functions to a halt, including malaria control. Land reform overturned rural class relations of owners and tillers, countryside and city. Food prices soared, housing supplies dried up, landowners cut down trees for fear that a new reform would claim those trees as state property (they were right). There were attempts, often brutal, to force famine-affected peasants onto new thinly populated land in lowland ecologies to develop in Ethiopia a new socialist model of development. Landscapes and local ecologies changed in willy-nilly patterns that were unintended consequences of politics and local upheavals. Among other things, these landscapes of revolution gave new life to malaria, which had receded markedly during the heydays of DDT spray and the Kennedy-era idealism of eradication.

Revolution meant forced movement and unsettled rural areas, areas of conflict, disrupted MES teams, cadres of students planted in resettled sites, and exposure to vectors/parasites. Best efforts failed as social systems broke down in the Blue Nile Basin and across the old empire.

One of the key elements of scientific socialism under the military government had been an aggressive and controversial plan called resettlement (*Aranguade Zamacha*, or Green Campaign), which included forcing farm families to move into villages (*mender serata*); and several student work campaigns, which sought to move famine-vulnerable highland populations to villages in lower-altitude zones that the government planners deemed appropriate for cash crops and agricultural mechanization. These initiatives were uniformly disasters. But the government's goal was to shatter the class/property foundations of that empire; the deeper effect was to reshuffle—disturb—the ecology in ways that invigorated disease ecologies. The mosquitoes loved it and moved into the vacuum.

In the 1974–1991 "long decade," the Malaria Eradication Service languished as a subunit within the Ministry of Health, quietly accepting a pragmatic policy of control and abandoning the idea of eradication as a practical goal, as the WHO had recommended back in 1969. With that, the MES faltered and died as the educated and activist staff came under suspicion by the military government of being too close to its rural constituency. The Jeeps fell into disrepair, lost their mission, and one by one were retired to the back lots of Ministry of Health stations in provincial towns, where their door decals faded and weeds invaded their gearboxes. The loss of trained, energized staff was substantially greater than the rust of the Jeep chassis. Malaria thrived.

From 1974 through the 1990s, Ethiopia's per capita aid donations from the international community had fallen to the lowest in the world. Malaria control was hardly a priority. In the socialist whirlwind of the late 1980s, Ethiopian health agents began to refer to the Malaria Control Agency, rather quietly dropping the reference to eradication. Things had come a long way since 1951, when the World Health Organization's H. M. S. Morin had officially reported that, thanks to the residual power of insecticides, malaria was definitely becoming a practically preventable disease.[24] After two decades of strategy and planning, the effort would change that prediction.

Dawn of a New Era?—1991

In May 1991, a triumphant rebel army moved into Addis Ababa at the end of a long march as a rural insurrection that had become a new government. The army of the former socialist government (the Derg) was in retreat and its political allies fleeing by any means necessary. Colonel Mengistu Haile Mariam, the head of state since 1975, boarded a secret flight to Harare, Zimbabwe, where he then cowered in exile in a small, protected villa in a neighborhood called, ironically, Gun Hill.

For its part, the World Health Organization had retreated but not given up. For a contract period of August 3 through September 18, 1991, it had funded a former MES employee and Michigan State University PhD in entomology to visit six districts of northern Ethiopia for an emergency evaluation of malaria epidemics that had broken out during the political upheaval. What he and other observers saw were roads in bad repair with burned-out Russian T-32 tanks and troop carriers strewn along the roadside. The hulks of those tanks and armored vehicles with government markings were all pointed south in their retreat to the capital. They bore spray-painted letters telling all who came upon them what unit of the rebel army had destroyed them.

The rout had been complete, with government buildings abandoned or occupied by young rebel fighters, often just teenagers, trying to rebuild some structure of governance. In the lakeside town of Bahir Dar, the once modern

Figure 4.8. A failed hope? Back lot of the Ministry of Health, Jimma (2003). (Photo by author.)

German-built hospital was a shambles with filing cabinets open, medical records strewn on the floor, and listless patients wandering in the walkways between buildings and wards. The few medical staff who remained looked drawn, overworked, and without lab equipment or space to meet patients or to diagnose or tend to disease, injury, or even childbirth.[25] Fuel stations in town sat empty; diesel fuel came on the black market, in one case from a drum beneath the bed of a bar woman. The infrastructure of daily life—and health—had fractured.

What entomologist Dr. Awash Tekla Haimanot found in September 1991, on his return to Ethiopia after two years in Geneva with the WHO, was disturbance. The collapse created a wonderland for mosquitoes and malaria parasites, but was of a profound nature for vulnerable rural people, who found the DDT-spraying program abandoned, staff dispersed by the civil war, and trained medical personnel having escaped across the border or to the outside world. Ecological conditions then had brought a very early outbreak of local malaria epidemics.

The renamed National Malaria Control Programme (note the abandonment of *Eradication*) was recovering from its own trauma. Dr. Awash reported a rather dismal state of affairs:

Figure 4.9. A failed hope. Abandoned MES vehicle. (Photo by author [2003].)

There was also a shortage of anti-malarial drugs, spraying of houses with residential insecticide did not progress as planned. While a reasonable number of localities were sprayed in Tigray [home area of the victorious rebel army] the operation in Gondor [*sic*] area was far below what was planned in that region. A total of 169 localities only were sprayed out of a total of a planned 821 localities. . . . Sectors are not yet operational as designated manpower and resources are not yet in place. The malaria control efforts are further constrained by shortage of vehicles and spare parts.[26]

Those Jeeps from 1964 had served long, but their days were over, as the mosquitoes frollicked and as the rains subsided, their density and malaria's virulence grew—a high vectorial capacity, as medical entomologists would say. Prime time for malaria. The chess game seemed over, and a dance of diving mosquitoes, parasites, and victims whirled about.

Chapter 5

Malaria Modern

1998, the Shivering Fever Reborn

> In late summer 1998 a severe and deadly malaria epidemic
> broke out in northwest Ethiopia. In the single district of Burie
> alone there were more than 42,000 cases and more than 740
> deaths, . . . 47 percent of the district's population reported hav-
> ing had the disease.
>
> —James C. McCann, *Maize and Grace*

In July 2005, Wolde Mikael found himself back in the trenches—
clad in rubber boots, gripping the handle of his plastic dipper, and search-
ing the surface of a muddy puddle for black dots. He found a cluster, deftly
scooped them up with a long plastic ladle, and poured the soupy mixture of
water, soil, and mosquito larvae into a plastic tray. He searched the puddle
again, and this time his practiced eye spied three wispy, inch-long bodies,
tiny heads on the water's surface and bodies draped at a 45-degree angle
beneath them. His shadow passed over them just as he reached the dipper
toward them, and in the blink of an eye they disappeared back into the
muck. A moment later they poked back at the surface, and, always patient,
Wolde Mikael used his eyedropper to snatch them into another plastic vial.
All in a day's work for the fifty-five-year-old, who had done the work of a
malaria field technician for forty years in and around the very site he was
working today. The thin bodies were the larvae of the *Anopheles arabiensis*
mosquito, and the black dots were pupae, both growing in the small, turbid
puddles and cattle hoofprints. The pupae were the last stage of mosquito
adolescence; they were soon to be adults that had enjoyed sufficient food in

the puddle to grow and then spread their wings as adult mosquitoes to seek sustenance—nectar for themselves and human blood for their eggs. They would emerge from the puddle into a dangerous adulthood.

Rockefeller Project Report, Summer 2005

In October 1998, the World Health Organization announced with considerable fanfare a new, ambitious program called Roll Back Malaria (RBM), which included powerful partners like the World Bank, the United Nations Development Program, and UNICEF. But there was also a strong sense of déjà vu, since this announcement followed almost thirty years after the WHO's 1969 sheepish admission that their pledge to eradicate malaria—and John F. Kennedy's 1962 dream—had failed.

In this same month that unveiled the ambitious Roll Back Malaria Program, the rainy season in Ethiopia ended with an angry return of that old nemesis— perhaps that region's most deadly malaria epidemic in recorded history. What happened? Why had four decades of antimalaria programs in Ethiopia left a population helpless? Was malaria in its modern form perhaps different from the episodes of the previous century? Remarkably, unlike the global broadcasts by BBC and NBC of heart-wrenching images of Ethiopia's 1984–1986 famine, in which people died in numbers similar to the 1998 malaria epidemic, Ethiopia's malaria crisis was a globally silent calamity. This chapter tells the local story of this epidemic that exemplifies the meaning of what we might call "malaria modern" in the Ethiopian setting—actually in two settings there.

Waktola's Story

On Friday, September 22, 1998, five-year-old Rosa Haji Hamed appeared with her father at noon inside the gate at the new Asandabo Health Center, a tin-roofed single-story building tucked into a eucalyptus thicket off a mud-caked road three hundred kilometers from Addis Ababa. Rosa had been experiencing a high fever, chills, and a headache for two days. The rains had subsided and the air was not so cold now. A few muddy puddles remained along the road to the clinic, reminders of the small pits, puddles, and pools of rainy season water that had been around Rosa's house the previous few weeks. Near the house, a few meters farther away, were a few broken pots and shallow pits in the white clay filled with water as opaque as strong coffee. Rosa's mother had dug those about two weeks earlier to mud-plaster the wall of their house. It had been a good maize harvest last year, and her father had built a small extension onto their neat thatched-roof house and garden.

Rosa's father, Haji Hamed, carried his child around the clinic building's veranda and approached the back side, where a metal-framed window stood open. They saw a crowd of people, mostly women, some with young children in arms, gathered around the open window. There were a few men also, looking peaked, wrapped in their white cotton *shamma* (togas), with their heads wrapped in cloths of various colors. A few wore baseball caps that said Baygon, or Pioneer, or had donned white *kofiya* skullcaps signifying their Muslim faith. One by one those in the crowd approached the window, stuck their right hand inside, and winced briefly. Then the next person approached, holding a feverous, barely responsive toddler. Her mother grimaced as she held the child with her arm outstretched and the man inside jabbed the child's right index finger with a needle and smeared her bloodied finger on a glass microscope slide.

Rosa's turn came eventually, and the man drew her blood with a quick finger prick and put the red drop on another glass slide. "Wait on the bench in the front," said the man in the Oromo language to Haji. Haji Hamed moved with his daughter in his arms back to the building's front side as the white-coated clinician directed. Inside the lab the man carefully placed the bloody slide on a wooden tray next to a weathered microscope on a table. He then turned to the next finger that appeared through the window casing.

An hour or so later the man called the name "Rosa Haji Hamed!" and then told Haji Hamed that Rosa's blood slide smear had shown *woba*—malaria. This meant that at some time in the past fourteen days, *Plasmodium* sporozoites had entered her bloodstream from the bite of an *Anopheles* mosquito. That day at the Asandabo clinic, 4 of every 10 patients tested positive for malaria. Later that month and the next month more cases would come, more children than adults, women and men adults equally, but overall in alarming numbers that exceeded living memory.

Back in Rosa's home village of Waktola, an hour's walk away along the new road to the east, the fields of tasseled maize surrounded the scattered houses of her neighbors and friends. Many of those people were also to suffer from the chills, sweating, and headaches. At first, many of them thought that the fevers were *tesibo* (relapsing fever), which was known to cause a sudden but temporary collapse of farmers—women and men—working in their fields, but that allowed them to recover eventually to resume their work, sometimes in a few minutes. But this disease was not as familiar. The clinic's laboratory, however, now told them that the affliction was *woba* (malaria), a disease they had encountered only rarely before. Woba was, as everyone knew, a disease of the lowlands, an affliction that others got when they traveled there. Why had this happened now, and what did it mean? Did Rosa's

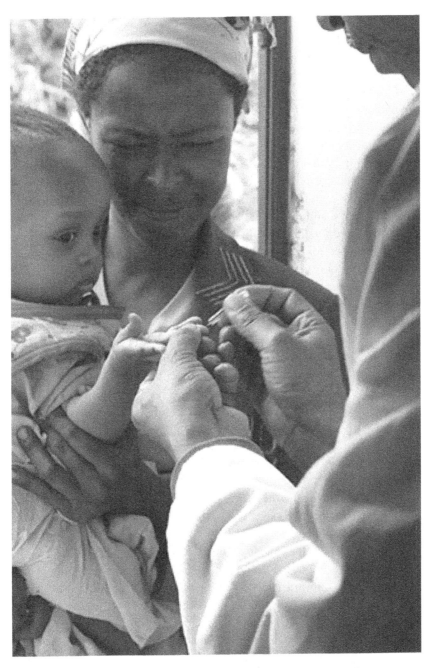

Figure 5.1. Blood smear and anxious mother at local clinic (2005). (Photo courtesy of Molly Williams.)

disease and suffering come by an accident of nature or was it a predictable, if not preventable event, or even a consequence of human folly?

Rosa's story was not an accident, nor was it a singular event, since malaria cases repeated themselves again and again across Ethiopia's mid-highland landscapes that October. In the case of the Waktola village's scattered houses, the malaria epidemic of 1998 foreshadowed a story of two landscapes, one of disease and the other of a new prosperity brought by agricultural change. And eventually the disease brought with it a collection of people: those who suffered from the disease's fevers, health workers who sought to cope with its persistence, researchers who wanted to understand why this sudden outbreak took place, and retired malaria workers needing an extra income from collecting larvae and replacing batteries on household mosquito traps. And then there were foreign specialists who came with measuring devices, free seed to negotiate for cooperation and help from farmers, and a hypothesis about how the disease and the agricultural landscapes had overlapped in such a deadly way. Did the spread of improved maize as a new crop bring on malaria in a new way?

Unraveling this deadly episode of malaria modern takes us from images at seven hundred kilometers in space to ground level — "ground truth," geographers and epidemiologists call it. The scale of life and images from a place called Waktola tell a story of malaria in the broadleaf forested zone of Ethiopia's southwest, where rainfall has been consistent historically and malaria only a sometime visitor and seasonal threat. Local folks knew malaria but saw it as a disease of the lowland "other" that they associated with uncultivated wetlands and river valleys (that eventually became hydroelectric dam reservoirs in the government's ambitious Gilgel Gibe hydroelectric scheme).

Malaria in Waktola has been an on-again, off-again thing — a part of life, but not enough for the bodies of Rosa and her family to gain "acquired" immunity from the disease as more often occurs in places in humid parts of West Africa where the infection comes every year and in every season. In those places infants often die, and those who survive carry a parasite "load" but show fewer outward effects of the infection. Infected persons carry a parasite load and thus can transfer the parasites, and the disease, to others via the mosquito's blood meal and her bite of the next victim. So Waktola and its ecology is what malaria folks/specialists call "unstable," meaning its on-again, off-again characteristic makes what looks to governments like a successful malaria control program but what actually plays out in daily life as an uncertainty and ends up with malaria still the number one illness listed on the charts of local clinic walls.

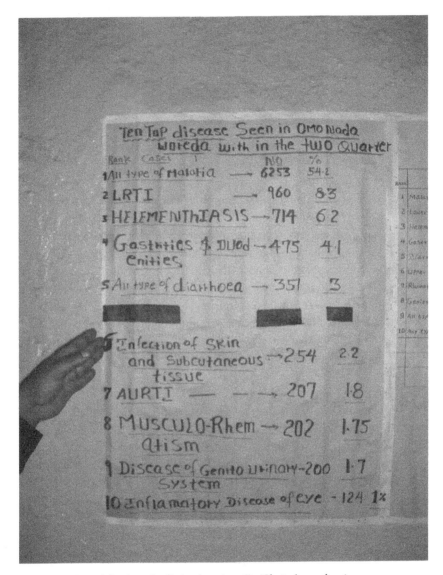

Figure 5.2. Asandabo (2005); clinic chart (2006). (Photo by author.)

Like Haji Hamid and Rosa, Waktola's people were lucky to have their clinic at a five-kilometer walk from their manicured compound and a thatched-roof home near their settlement's modest, single-room mosque.

We have very few records or historical descriptions of malaria in the Waktola area. Farmers I worked with and interviewed there recalled malaria as a disease of the lower zones orf the Gibe River valley several kilometers away. There is a bit of older evidence, too. In the nearest town, Jimma (fifty-five kilometers to the west), on a short 1957 visit, Sir Gordon Covell, the WHO malaria specialist, offered a snapshot of malaria using "spleen rate" (the count of palpably enlarged spleens) exams of children and local ecology in a year with no local epidemic:

> Children were examined at three different schools in Jimma, total-
> ing 234. The combined spleen rate was 8.6 percent, the majority of
> spleens recorded as enlarged being only just palpable. Replies to
> local inquiries indicated that malarial incidence had been slight in
> recent years, and this was attributed in some quarters to the plant-
> ing of large numbers of eucalyptus trees and improved drainage. An
> Italian physician in charge of the local hospital said there had been
> a serious epidemic some years ago, he thought in 1944. It seems
> that as in so many other areas of Ethiopia this was an area subject
> to periodical regional epidemics and judging from the present low
> spleen rate and consequent lack of immunity, a severe epidemic is
> to be expected whenever conditions become especially favourable
> for the production and longevity of the local mosquito vector.[1]

Covell's report gave us a simple snapshot and two important facts. First, the low "spleen rate" meant that infants and toddlers in the Jimma area had had only limited exposure to malaria infection—and no acquired immunity. Second, local memory tied the low rate of malaria in the recent past to a subtle ecological change. Those local folks had their own sense of the effect of ecological change: It was eucalyptus tree plantings that had lowered the water tables and drainage associated with mosquitoes along with the urban growth of Jimma town.[2]

This was their attempt to explain malaria's changing personality. Malaria was or had become "unstable." And, overall, Covell and people he met on the ground linked malaria's epidemic patterns to the fluctuations in mos-quito populations and the ecological conditions that made those numbers grow or recede as local ecologies changed seasonally or as human victims, aka malaria reservoirs, moved about. Spleen rates were a crude measure in-deed, but there were no laboratories in 1957, and pressing the distended rib cages of children offered a palpable glimpse of malaria's past paths.

Let's now revisit 1998, but move north and back to the Blue Nile water-shed and very near the site of James Bruce's 1770 visit to the symbolic source of that river. Here is where, in October 1998, in the county-sized district of

Burie, malaria swept outward from the western lowlands onto the densely settled highlands that had undergone an agricultural transformation akin to Waktola's. In an area that had not experienced malaria in modern memory, a malaria tsunami surged and spilled onto the highland settlements.

Burie: Malaria in the Heart of the Watershed

In late summer 1998, a severe and deadly malaria epidemic broke out in northwest Ethiopia as it had in Waktola, seven hundred kilometers by road to the south and west. In the single district of Burie alone, there were more than 42,000 cases and over 740 deaths, meaning 47 percent of the district's population reported having the disease in the period from June to December. But the outbreak was not evenly distributed. In one village the local health official closed ninety houses, the residents having died from a lethal falciparum malaria. In a local school in another location, half the children lost one or both parents. Yet, in another village at a similar altitude, there were few if any cases at all. Some places that had had no cases during September then had hundreds in November. The patchwork quilt of infection brought confusion and desperation. As with our 1953 narrator Asres, Burie's shell-shocked farmers blamed *zar* spirits, consulted diviners, and sought propitiation by slaughtering black oxen; others sought succor from the few government health stations scattered within the district. The new Health Center at Burie town had the area's only laboratory.

This spotty and bewildering "hot zone" outbreak continued through the fall months and gradually dissipated by January 1999. What made this outbreak the more puzzling was that it affected a region that had never known epidemic malaria before in collective historical memory. Indeed, in the mid-1970s there had been neither mosquitoes nor malaria.[3]

What accounted for this new infection, its devastating intensity, and its uneven distribution across the rural landscape? When asked in 2005, residents and malaria researchers I spoke to had no immediate answer to that question from either residents or regional health officials. Some residents recalled unusually intense indoor mosquito biting activity that August and September. But that evidence was anecdotal. In 1998, local Orthodox Christian rural folk in some districts had blamed *zar*, the evil spirits, and tried to placate them. Others, like their Muslim fellow farmers in Waktola, called it *tesibo* (typhus) and avoided contact with their neighbors for fear of infection. Few, if any, local residents associated this deadly outbreak with the *nidad* or *enqetqet*, two local names for malaria, a disease that they associated with distant lowland areas and not their highland plateau.[4]

New research, however, raised a novel question of whether there was a relationship between this new intensity and the geography of malaria in northwest Ethiopia, and a new agroecology of maize planting that enveloped the area at the same time. Addressing this question of disease links to agriculture and farmers' crop choices required the detective skills of an environmental historian, an epidemiologist from the Ministry of Health, and a young entomologist, all who shared a deep interest in both the human tragedy and the environmental puzzle.[5] Part of the solution goes back to a hunch pursued by Yemane Ye-Ebiyo, a young Ethiopian entomologist engaged in a PhD program at the Harvard School of Public Health in Boston at about the same time as the epidemic engulfed northwest Ethiopia. What follows now offers a lesson in research design and its challenges in a dynamic ecosystem where the parts are always shifting.

Yemane's idea had begun in Ethiopia and concerned his and his Harvard adviser's curiosity about the feeding habits of the larvae of *Anopheles arabiensis*, the mosquito that has been and is the main vector for malaria in Ethiopia. Yemane and Dr. Andrew Spielman's question was: What did those larvae eat in those small puddles and pits that allowed them to survive into adolescence (as pupae) and adulthood as blood-hungry adults? In short, they were specifically interested in what effect the food supply had on the survival and growth of mosquito larvae and how successful the larvae were in advancing to the pupa stage of their life cycle. To test this idea, Yemane set up a controlled field experiment near the town of Ziway in Ethiopia's Rift Valley to see the effect of plant pollen on the mosquitoes' breeding cycle. For his experiment he chose maize, that area's most commonly grown domestic plant and increasingly Ethiopia's major food crop. The Yemane/Spielman hypothesis was that since maize was one of Africa's few wind-pollinated field crops (as opposed to self-pollinating or insect-pollinated crops like wheat and teff), maize pollen might indeed fall as food on the local turbid water that was the mosquitoes' favored breeding site. What, he asked, was the value of maize pollen as a food source for the mosquito larvae?

Yemane's field experiments compared breeding sites that included maize pollen and those that did not. Cleverly, he also compared sites that were within ten meters of a maize field and those at least fifty meters away; wind carried pollen dust to distant puddles and pits. Overall, he compared the effect of maize pollen on the number of larvae that survived to the pupa stage, and thus to blood-hungry adulthood, the speed of this growth, as well as the size of the adults that emerged from the breeding sites, and the females who were at the heart of the story, carrying parasites to humans in their bites and then laying eggs.

His results were astonishing. Did the maize pollen that dropped into puddles, pits, and hoofprints affect growth of the larvae? Of those larvae that had maize pollen as a nearby food source, virtually all (94.1 percent) survived and developed, advancing to the adolescent (pupa) stage; in comparison, just 13 percent of those located at more than fifty meters distance from the two-meter-tall maize plants survived to the pupa stage. Eureka.

In a parallel experiment, Yemane compared breeding sites sprinkled daily with maize pollen with those under local conditions but with maize pollen excluded. The result was that maize pollen–fed larvae almost invariably developed to pupae more than ten times more often.[6] What was even more telling was that the size of the maize-fed adults was almost uniformly 13 percent larger than those deprived of maize pollen. This was a critical factor, since earlier work had already established that the larger the adult, the longer the life span, and thus the greater possibility of a dense mosquito population with more opportunities to carry parasites from one human victim to another.[7] Yemane and his colleagues were left with this question: Was maize an integral part of the human/mosquito landscape that fostered malaria? And could this maize-malaria linkage be proven in an actual epidemic situation?

With these data in mind, I approached Andy Spielman, then a professor at the Harvard School of Public Health and coauthor of Yemane's PhD paper, about the possibility of testing those findings in an actual agroecological setting. I was particularly intrigued in pursuing this work in northwest Ethiopia's Burie district, an area at the heart of the 1998 epidemic and where I had lived for two years in the mid-1970s.[8] I had the tail end of a Fulbright-Hays grant to work in Ethiopia, though Yemane's Harvard work meant he could not leave Cambridge. The Harvard team then put me in contact with Asnakew Kebede, an epidemiologist in Addis Ababa with the Ethiopian Ministry of Health, who had worked for ten years in the Burie district. After comparing our field experience and ideas on malaria and its ecology, Asnakew and I resolved to test the hypothesis that the intensity and spatial dimensions of the 1998 epidemic were a product of several coincidental factors, including the new agroecology of maize that contributed to a disruption of a long-standing disease equilibrium that had held malaria in check at the research site until the recent and sudden 1998 outbreak.

We also learned another fact that seemed to matter. As a wind-pollinated crop, maize massively overproduces the pollen it needs to cross-fertilize nearby plants. One maize plant can shed as many as fifty million pollen grains in a single season; one densely planted hectare yields 150 kilograms of pollen per season. This was an amazing part of the malaria equation.

Figure 5.3. Recording malaria data in the Burie Health Center (2003). (Photo by author.)

Maize pollen grains are nearly round and about as wide as a human hair. They are light enough to be carried by the wind—but not too far. Maize pollen will fall to the ground within about thirty meters of the plant that produced them. Grains that fall to the ground may also be washed into nearby puddles, dusting the puddle surfaces with nutritious organic particles—food for mosquito larvae.

With this in mind, we needed to rethink the full effect of maize as Ethiopia's increasingly dominant crop. Was maize feeding mosquitoes as well as people? The next step was to put the pieces of the malaria-maize puzzle together to reveal how this crop affects the real world of farmers' lives and malaria's dance. In May 2002, we traveled to Burie and began to collect handwritten health records from local government health centers and clinics as well as agricultural records from the local Ministry of Agriculture offices. We spent long days and long nights with pencils and waning laptop batteries. We also gathered statistics and qualitative information on agriculture and environment from various national research institutes on crop production and walked over farm landscapes with helpful local farmers identifying crops and human settlements.

Figure 5.4. Maize field habitat; ox-plow with farmer. (Photo by Molly Williams.)

We collected farmers' ideas and oral tradition from local folks affected by both the malaria epidemic and the emergent agroecology of maize. As we traveled to the malaria-stricken agricultural areas outside of Burie town, we inspected farmsteads for mosquito breeding sites, proximity to sources of maize pollen, and human parasite reservoir populations.[9] There were one-on-one interviews in farmhouses and group interviews hosted with local maize beer (*talla*).

We wanted to understand the logic of the farm ecologies themselves. Like Waktola's, Burie's historical highland setting is a subset of the classic Ethiopian highland ox-plow agroecology. Burie followed a classic Christian pattern of small ox-plow farms. Waktola was an Islamic community of farms that also used the ox-plow, managed their soils for grains, and had adopted late-maturing maize—Muslim farmers in Waktola and Christian farmers in Burie alike. Both farming systems have made their historical footprint within a summer-wet/winter-dry rainfall climate and annual cropping of endemic cereals (teff and finger millet), as well as grains from Ethiopia's unique niche as one of Vavilov's centers of secondary diffusion (wheat and barley), plus pulses (chickpeas, horsebeans, field peas) and oilseeds (nug, safflower). This was a farming landscape beautifully adapted to soils and rainfall, and the ancient, efficient technology of the ox-plow.

Agriculture was old, but malaria was unexpected by farmers who worked the highlands. Burie residents' folk understandings of malaria's symptoms, cause, place of origin, and prevention were a product of their limited exposure to it. Many of the above-fifty generation of Burie residents recalled the disease as exclusively a product of lowlands and marshes. *Woba* was the word used in modern Ethiopia and the ministries. *Nidad* (literally, to catch fire), as Burie folk called malaria, was a disease of intermittent high fever that afflicted people who passed through the Blue Nile watershed lowlands, especially Burie's Muslim merchants, who did not farm but regularly traversed the lowlands between the major market centers in the empire.[10] Rarely fatal, *nidad* affected travelers after their return home, but the absence of the mosquito vector in the higher zones, which most people occupied, meant there was little if any transmission on the highlands themselves. Malaria was, after all, a sometime visitor to higher altitudes, coming once or twice a decade. Another name for malaria recognized locally was the onomatopoetic *enqetqet*, the "shivering fever" that sometimes affected those who slept outdoors alongside their herds at the edge of the low-lying unpopulated wetlands—a phenomenon that changed as population growth pushed farmers onto lower areas. But the most provocative association of malaria with agriculture was a ubiquitous folk adage that putatively linked malaria symptoms to maize. One informant recalled, with some amusement: "Our fathers used to tell us that to frighten us. 'Kids! Do not cut or ruin the tassels or eat the stalk of the corn or you will get shivers [*enqetqet*].' Our fathers said this to make us afraid. They said this to frighten us. The shivering fever."[11]

Most likely, this popular expression was their connecting the growth of the temptingly sweet maize stalks, the need to protect the valuable green ears from mischievous children, and the appearance of malaria in late August and early September. It was then when the force of the rains lessened and nighttime temperatures rose above the key 15–18 degrees Celsius threshold necessary for parasite growth in the mosquitoes' gut. Was it intuition or long-lived experience that the cultural memory of Burie folk (and those in Waktola as well) made the association between malaria season and the maize plant life cycle? Overall, however, the historical malaria-free agroecology of Burie offered a reasonably stable balance between only mildly endemic malarial ecologies off in the lowlands, a very limited mosquito population, the absence of a local human parasite reservoir, and ambient temperatures less than 18 degrees Celsius (65 degrees Fahrenheit).

Burie's agriculture and its relationship with malaria had evolved over many generations of smallholder agriculture and Burie's regional role in trade. Gazing out at the agrarian landscapes that surrounded Burie town in the 1970s, one could easily witness persistent practices that gave its fields and landscapes their visual character.

The end result of the plow, farmers' local experience, and crop mixing was the patchwork-quilt landscape of small plots growing diverse crops. From my own observations while living in the area for two years in the mid-1970s, I had in my mind's eye a human landscape of dispersed homesteads surrounded by fenced gardens and crop fields scattered willy-nilly across the rural scene—green cropped fields in the summer and golden at harvest in the late fall; brown harvested fields by December and January. This image hardly seemed to support the formula for a maize-malaria link in terms of the proximity of maize field to mosquito breeding sites around the house. I suspected that the maize-malaria hypothesis that came from Yemane's work on pollen and larvae would come unglued on that issue alone. I was wrong about that.

Burie's farms sit on a mid-altitude plateau 2,000–2,300 meters high above a marginal lowland descending southward into the Abbay (Blue Nile) gorge, where the population is sparsely settled, mainly along the new road that crosses the Blue Nile River. That road retraces the old caravan route that linked the coffee-rich areas on the Blue Nile's southern bank to Burie, and thence to the old imperial capital at Gondar and destinations north. In those low-lying, thinly populated zones, malaria is endemic, and the parasites in people with acquired immunity there make it a silent reservoir of malaria that springs forward periodically—and unpredictably.

The local agricultural system, as it turns out, had become a part of this dance. Burie farmers, like their compatriots in Waktola, used little chemical fertilizer prior to 1980, though they had long applied manure and ash to household garden crops such as maize, kale, capsicum (red pepper), and herbs cultivated inside their fenced household garden plots, called *gwaro maret*. In this old farm landscape, maize plants were a minor partner: local, early maturing types germinated with the first rains that shed their pollen in the cool months of June and July when heavy rainfall washed pollen out of the fields, and the plants produced green ears in August and September. Roasted or boiled, these young ears (Amharic: *eshet*) offered welcome snacks when food shortages preceded the main harvests in November and early December. But early maize types did not coincide with mosquito egg laying and the appearance of larvae in the puddles. That was to change with the arrival of modern maize and modern agricultural practice.

The Changing Agroecology of Maize in Burie

Until the 1980s, maize was a minor field crop in the northern highlands and appeared on farms mostly as a garden vegetable consumed green in the "hungry season" (August–September). Waktola was similar, although sorghum was also a familiar crop in the adjacent malaria-prone lowland areas. Farmers chose from among an array of early maturing local maize types; the best-known ones they called *Mareysa, Harer,* and *Kafa*—names that suggest an origin in southern Ethiopia. Some of these probably arrived in Ethiopia via the Nile Valley in the nineteenth century, and some came from U.S. agricultural aid programs of the early 1950s, while others may have been descended from the original imports brought from the New World (via India) by Arab and south Asian merchants plying the Red Sea trade in the sixteenth century. Maize's most common name in northern Ethiopia was *YaBahar mashela* (sorghum from [across] the sea). At first, maize had a value mainly as a welcome snack for farms awaiting November–December cereal harvests. By the time malaria season came to the lowlands below, the maize on plateau fields near Burie was already ripening and fallen pollen had washed away with the heavy rains of July.

As noted above, malaria loves disturbance, in both ecology and politics. In the 1980s and 1990s, Ethiopia had both. In the early 1980s political changes brought a transformation in Ethiopia's agrarian economy and agroecological balance. Under the socialist military government known as the Derg that ruled from 1974 to 1991, grain marketing policy, forced labor, insecurity over landholdings, and government food security programs brought increasing political chaos but also changes in Ethiopia's national crop patterns. Ethiopia's socialist government saw maize as a high-yielding field crop to replace labor-intensive teff, a disease-prone wheat, and poor-yielding sorghum. As in the Soviet Union, Ethiopia's socialist planners used maize as the ultimate product of an industrialized, scientific agriculture. For their part, farmers saw maize as a low-labor, quicker-maturing crop that provided food in insecure times when the socialist state forced their labor onto public works projects like tree planting or fixed the prices for their other farm produce. Farmers were confused and maize was a rational choice in troubled times. By the mid-1980s maize had unceremoniously surpassed teff and barley as the major grain crop produced in Ethiopia. Maize directly superseded sorghum on low- to mid-altitude fields, replaced coffee in some areas of the south, and complemented fields of the narcotic leaf *chat* in the southeast. Maize was the primary focus of state farms in the southern and western parts of the country.[12] By the mid-1980s, improved open-pollinated maize had entered Ethiopia's northern agrarian economy as a major field crop.

In 1995 (after the 1991 fall of the socialist military government), the Ministry of Agriculture and Jimmy Carter's nongovernmental organization Sasakawa Global 2000 brought an infatuation with new hybrid types of maize and began a demonstration package program to expand the use of inorganic fertilizers (heavy on nitrogen), improved maize seeds, and agronomic techniques such as early row planting and intense early weeding to improve national food production. It worked. By 1998, nationally maize had reached 32.6 percent of the country's cereals production, a percentage higher than any other grains. Between 1993 and 1998, maize's area of cultivation increased 79 percent (from 808,900 to 1.45 million hectares). In the northwest Amhara region (where Burie is a district), adoption was higher even than the national rate.[13]

The chief catalyst in the agroecological metamorphosis appears to have been that new and enormously productive hybrid maize unromantically named Bako Hybrid 660 (or BH660, locally called "Silsa Sidist" [i.e., Sixty-six]). For the farmers the new maize's height and heavy stalks were a thing of beauty. The variety produced prodigious yields of six to eight tons per hectare when planted in red clay soils and with consistent months of rain in the growing season. This was true for both Waktola and Burie. Released by the

Ministry of Agriculture in 1993 and introduced to the Burie area beginning in 1995, by 1998, BH660 had replaced virtually all other varieties in both areas. A 1998 survey of the entire Amhara region indicated that 80 percent of the farmers sampled had adopted improved maize, with BH660 as the most popular variety overall.[14]

Maize's remarkable expansion within Ethiopia's farming systems over the final decade and a half of the twentieth century had in many areas amounted to an agroecological sea change in labor calendars, affording an opening to national markets and newfangled seed, nitrogen-heavy fertilizer, and storage places to keep the surplus. This new wealth increased reasons for farmers' passion to plant more hectares in maize on as much land as possible. The result was a bold new landscape that changed the countryside and created new cropland for maize that came right to the doorsteps of farmhouses. Maize ran amok. It was a new landscape, one that perhaps had unexpected consequences for health.

For 1998 there was to be a grand transformation. While there were not dramatic increases in temperature, somewhat heavier rains in 1998 did not demonstrably lower ambient nighttime lows or daytime highs. This was a year of consistent rainfall that extended well into the fall months. Overall, neither temperature nor rainfall fluctuations would seem in and of themselves to account for the precipitous shift in malaria cases, as climate records show no unusual events of either rainfall or temperature.[15]

Maize's spread as the crop of choice in the Burie area during the nine years after the 1995 introduction of the hybrid maize package was remarkable, more than doubling the area planted to maize and raising the percentage of the total area devoted to maize from 21 to 36 percent.[16] While these local figures mirror national trends toward maize's domination of cereal production and document its local face in Burie, they mask the even greater dominance within Burie. The final figure of 26,758 hectares planted in maize for the 2001 crop year indicates that in Burie itself maize accounted for well over a third of all cereals, and was growing rapidly overall. For the 1998 crop year, it was certainly at least that high: Maize's share of total cereal planted probably reached 85 percent.[17] One farmer commented to me that "if a thief hides in a maize field beginning in Gulim, he can run under cover of that maize until he comes out at Fetam Sentom [about twenty kilometers to the south]."[18]

Burie farmers' heavy concentration in maize cultivation brought with it a significant change in the area's pattern of human settlement, something we might call "maize high modernism."[19] Those places most intensely invested in maize farmers had changed their patterns of labor, fuel supply, livestock forage, and domestic space. Maize madness. Unlike former dispersed

Figure 5.5. Maize fields prepared for planting at Kuch health station. (Photo by author.)

homestead sites, where fenced garden space separated houses from field crops, farmers now had clustered homesteads within small nucleated villages/towns of corrugated-iron roofs and rectangular houses. Those settlements included grain storage buildings, teahouses, shops—all signs of the increasing income base of rural dwellers. Most remarkably, the houses in these villages have no defined *gwaro* (garden space), with maize cultivation running up against the walls of the houses themselves. Some plantings stood within a half meter of the houses, and virtually all of the dwellings we surveyed had maize fields within ten meters of the dwelling and most much closer (i.e., well within wind pollen movement). When I asked a farmer about the disappearance of his fenced house garden, he laughed and said, "You see that stack of straw there [adjacent to a house]? Once the cattle finish that one, we are going to plow up that land [for maize] too!"[20] The pits and puddles near the houses also nurtured larvae and were perfect landing places for the falling windblown pollen.

The physical appearance of these new villages in maize locales gives a spatial expression to a deeper shift in material culture. Farmers in the area insisted with enthusiasm that maize touched all aspects of their lives, and for both genders. Maize stalks provided cooking fuel, green leaves were live-stock fodder, young ears offered snack food, and women now mixed maize flour into *injera* batter (Ethiopia's distinctive teff pancake) and into wheat bread. Shelled cobs appear around households in myriad uses, and often just piled in corners. Instead of the ubiquitous stacks of teff straw as dry-season livestock fodder, virtually all houses in these areas sport a lean-to stacked neatly with dried maize stalks preserved for fuel. Maize kernels roasted, ground, and fermented make quite excellent local beer. Pre-cious teff straw, so loved by farmers and their cattle, is now less a farm by-product than a commodity purchased from the market with profits from bulk maize sales, an interesting measure of agricultural development. Pros-perity brought new house construction, or house repair using plaster made from clay dug from shallow clay pits around houses and near fields.

The maize type and its specific characteristics contributed directly to the agroecological change that underlay the 1998 epidemic. Most important to the local malaria equation, maize's tasseling and showers of pollen release takes place later in the season than previous types of grain (i.e., late August–September, at precisely the time when temperature and moisture are ideal for mosquito breeding). This confluence in the agroecological calendar had serious implications for malaria infection in a modernizing ecology. Above all, it is maize's prodigious yield of grain that prompted market-sensitive farmers to sow it in every available space, even in school yards and in the fenced compounds of health centers. Recall that amazing fact: One hectare of maize produces 150 kilograms of pollen. Wheat, barley, rice, and sor-ghum are self-pollinating and produce only small amounts of their pollen.

The 1998 Malaria Epidemic in Burie

The number of malaria cases and deaths in 1998 was unprecedented in the agroecologies of both Burie and Waktola. Previous disastrous malaria epi-demics in Ethiopia (in 1953 and in 1958) had affected wide areas of the coun-try and the Lake Tana region, but did not reach Burie. In 1998, by contrast, in the Amhara region alone 3.4 million people were affected and more than 7,700 deaths occurred.[21] The outbreak in the Burie area began in thinly populated lowland endemic areas in May and June but then expanded into higher-altitude zones in September–October 1998, and exploded onto fur-ther new ground in October and November when the powerful wave of ma-laria washed onto previously malaria-free zones. It was as though someone had poured incendiary fuel onto the epidemic fire. The outbreaks finally subsided in January 1999, leaving exhausted victims and officials to wonder about its causes and consequences. The dramatic August–September 1998 spike in malaria infection coincided with the BH660's pollen shedding and mosquito larva hatching into pollen-rich puddles and pits, suggesting a connection between those two events, a link now borne out by Yemane's field and laboratory evidence.

But how much of this ecological change and the malaria outbreak was mere coincidence? Nature—natural systems—rarely allow nice, neat, and single-stranded cause/effects. Where did maize fit into this off-and-on, swirling malaria dance?

Early findings on the relationship between maize and *Anopheles arabiensis* in controlled experiments raised the question of whether actual ecologies on the farm, rural settings, and the late-season sudden arrival of maize produced an increase in what malariologists call the force of transmission, providing key components that underlay the virulent malaria firestorm. The 1998 malaria epidemic in Burie and Waktola offered compelling reasons for investigating the role of maize in malaria transmission and to add to the history of malaria overall. The timing of pollen release, the presence of new breeding sites for larvae, and larger and longer-living adult females provided a potentially deadly formula—a perfect storm—for an ecology of modern malaria. The shedding of pollen close to human housing strongly suggests that the new agroecology of maize and malaria supported increasing mosquito density, an increase in the mosquito's breeding space, and an increase in the rate of her bites—these factors all leading to an outbreak of epidemic proportions in the malaria tsunami of 1998. In the nutrition-rich ecology of monocropped maize, mosquitoes live longer and bite more times, given those longer lives, as they pass on their malaria parasites.[22]

Let's consider two lines of evidence. The first is the statistical association between the presence of maize and the chronological and spatial impact of the disease in those summer months of 1998. What is most remarkable is the case data from local clinics and health centers from August to December in 1998. Second is an assessment by the Ministry of Agriculture's Burie district office that ranked agricultural locations by high, medium, and low density of maize cultivation.[23] When the factor of altitude is held constant (since that affects temperature), the statistical impact of heavy maize cultivation on malaria infection is shocking. In areas of high maize production with the concomitant agroecology of proximity of pollen sources and mosquito breeding sites, 24 percent of the population registered positive diagnoses of malaria (*P. falciparum* and *P. vivax*). In areas of medium maize production, 9.2 percent came down with the disease. And in areas where maize was only sparsely cultivated, only 2.5 percent of the population contracted malaria. The ratio of cases per population in high versus low maize areas was 9.5 to 1.0. In other words, farmers who grew maize in Burie's densely cultivated mid-altitude zones were approximately ten times more likely to contract malaria than those who cultivated other crops in the same altitude. Wow.

Modern Malaria's Complexity: Rockefeller Results

The question is how to study modern malaria in a world where the pace of the malaria dance is accelerating: people moving to cities with new mosquito habitats, new ecologies of peri-urban agriculture on the edges of those cities—new watery habitats appearing haphazardly, and human parasite reservoirs on the move in new rural economies.

In 2005, the Rockefeller Foundation decided to support our team of entomologists, geographers, and an environmental historian to work with Ethiopian counterparts to study this maize-malaria puzzle. Or is it a conundrum? After all, the maize-malaria work connected two of Rockefeller's primary goals for the past seventy-five years: to control malaria and to raise world food production by improving maize as a world food source. The 2005 team's goal was to investigate the role of maize cultivation in increasing malaria transmission. As Andy Spielman and I left the Rockefeller office headed for the airport after our first meeting, Dr. Gary Toenniessen, the accomplished microbiologist and Rockefeller program officer, insightfully gave the team a simple goal: "Prove yourselves wrong!"

The maize-malaria project goals included adding a novel, ecological view of malaria's persistence and dynamism over time. But how might maize's rapid rise as the crop of choice for small farmers connect to malaria's resurgence? Biomedicine's malaria failures to date have, in fact, begged a return to a fuller ecological understanding of the ecology of malaria. The role of intensive maize cultivation in malaria transmission is a part of this complex ecology. We need to rejoin the struggle by paraphrasing a political metaphor, that all malaria is local and is a complex tapestry of nature's forces.

After five years of often difficult field and laboratory work, peering at satellite imagery and working with teams of farmers and local clinics, in fall 2011 the Boston University–Harvard School of Public Health team submitted a report on five years of field study.[24] The central goals and findings are worthy of the following summary as insights into malaria and its changing Ethiopian ecology.

Our ideas came orginally from a series of experimental observations that implicated maize pollen as a nutritionally important source for the larval stages of the *Anopheles gambiae* complex of mosquitoes, the main vector of malaria in sub-Saharan Africa. But the approach also came from observing that malaria transmission connects directly with the intensity of maize culture in a region in Ethiopia where transmission is "unstable."

Trends in maize food production in sub-Saharan Africa supported the suggestion that people living and working on or near smallholder farms that increasingly exploit maize production may also increasingly bear the burden of malaria. The Rockefeller Foundation's commitments to improved agricultural productivity and the betterment of human health made sense for their support of an intensive study of the potential unintended consequences of the push for maize as an engine for global food supply.

We sought to measure the way maize cultivation contributes to the increasing burden of malaria in sub-Saharan Africa. But that continental scale misses the important ecological nuances of place and people. We chose the local scale, the sites of Waktola and Burie, to measure effects that were laid out in numbers and laboratory results—PCR, wing length, larval feeding—but also in human terms among the farm families in those places.[25] Toward this end, we devised an array of standard and novel field and laboratory studies. In particular, we markedly limited the shedding of pollen by removing the dusty yellow maize pollen in certain fields of maize during the malaria transmission season in selected areas of a village in which this infection is endemic and sought to determine the extent to which we could reduce diverse components that contributed to the force of transmission.

Designing the Study

To test the hypothesis about maize pollen in the complex realities of farm fields, we tried "ablation rescue" experiments, where we reduced the source of an effect, pollen in this case, recorded the results, and later put it back. We removed the pollen-bearing tassels in stands of maize within fifty meters of a defined major *Anopheles* mosquito developmental habitat and hand-pollinated them to simulate a normal year of growth. For comparison, we examined an ecologically similar set of maize fields adjacent to other larva habitats. We converted a third set of fields, at the choice of the farmer, to other crops—such as red pepper, taro root, teff—that did not shed wind-borne pollen. We alternated the study blocks each year for five years as part of the "rescue" portion of the study, and we anticipated that these replicated paired observations would help resolve the link between maize cultivation on small farms and malaria risk. There was much to be learned and many tweaks in the methods needed to get results that fit the real world.

We then chose an array of farms on which the BH660 late-maturing variety of maize was simultaneously planted and cultivated in a fairly uniform manner. Each farm included several cultivated plots, one or more defined mosquito habitats, and the modest mud-plastered buildings used by the farmers' families for domesticated animals and storage of the livestock feed.[26]

The results of our efforts will undoubtedly inform agricultural and public health decision makers regarding the intricacies of the agroecology of maize and the influence of its husbandry on vector-borne infections.

Larva Effects

The larval stages of *Anopheles arabiensis*, the major vector of malaria in the more arid parts of Africa, particularly thrive in turbid, vegetation-free puddles of water. Borrow pits, the depressions that result from the digging of fine clay used as plaster in traditional mud-and-wattle houses and huts, serve as particularly suitable habitats for mosquitoes. These pits are usually near human settlement sites: The structures in which people sleep and structures used to retain water from rain and runoff support an impressive and diverse array of plants and animals around house sites. People use those water pits for drinking water for their domestic animals, for bathing, as places for children's play, and as convenient pools for washing clothes. We might expect that the prominent inorganic and disgestible food matter in such water would seem to deny these larvae the available organic nutrients needed for their development. We found, however, that the midguts of these larvae in nature frequently are filled with maize pollen and their development speeds up when the watery habitat is near flowering maize pollen. The pollen stimulates their intense feeding.

These observations suggested to us that maize pollen may play a supporting role in the transmission of malaria near small farmsteads. The intensity of maize cultivation near homes matches up with malaria risk during epidemics like the one in 1998. This needed a closer look, as these findings would hold important implications for public health, agricultural extension, food security, and health equity in rural communities in Ethiopia, Africa, and beyond.

We found the apparent maize-malaria link to be a surprising and unanticipated consequence of the agroecological dynamics of that dramatic expansion of maize production in Ethiopia during the past several decades. The swirl of malaria's historical dance that appears, turns away, and then reappears when ecologies change may well respond to this new crop ecology or others like

Figure 5.6. Maize-malaria field site work (detasseling). (Photo by Richard Pollack.)

it. Maize's expansion in Africa and Asia is a function of concerted efforts by national and multilateral policy emphasizing high-yielding, improved maize for human food supply in Africa and urban meat production in Asia. It is an example of what may lie ahead in a world of complex ecologies of disease.

Africa's maize geography significantly places poor households at a disproportionate level of risk for malaria relative to larger-scale maize farms or farmers of other crops. The dilemma is that small farms at the opening of the twenty-first century are the predominant source of maize for national food supplies in most of southern and eastern Africa. The maize breeding philosophy of the 1960–1995 high-yield hybrid era assumed that small farmers would buy an increasing amount of new seed in the ideal settings of "high-potential" areas for fertility, rainfall, and economies of scale. The small farms that increasingly predominate in African maize production necessarily mean that the most productive habitat for malaria vector *Anopheles arabiensis* is near a potential key accelerant (maize pollen). African small-farm families necessarily live close to their maize fields and so face a level of risk significantly higher than do commercial farmers. What is even more important, the new "modern" maize varieties are longer maturing and shed pollen later, closely matching the breeding season for the mosquito vector in rain-fed cereal agriculture.

During the first project year, we chose the farming community of Waktola, fifty-five kilometers from the regional capital of Jimma, a place that Sir Gordon Covell had visited and examined children's spleens in 1957. Our first step was to map the area using GIS satellite images, with particular attention to potential anopheline breeding sites and to the distribution of gardens and nearby agricultural fields. We wanted images of fields and crops to compare to what we actually saw on the ground during our work with farmers on what, in 1998, had become a landscape of epidemic malaria. We could easily see predominant larval breeding sites, cultivated fields, homes, and roads/paths from the satellite images, which we confirmed by ground truth surveys of what the land looked like at ground zero as we walked through fields, stepped over muddy hoofprints, and talked to farmers about cropland use and soil types. This made for lively conversations with farmers, laughter, and debates among the team.

We sought to see the shifting array of elements in local "forces of transmission" of malaria, including entomological inoculation rates (EIR)—mosquito bodies or indices (called EII)—with which we could compare treatment sites through time. We sampled adult mosquitoes using standard CDC light traps to estimate adult population levels. We wanted to see the number of adult mosquitoes, since mosquito population was an exponential

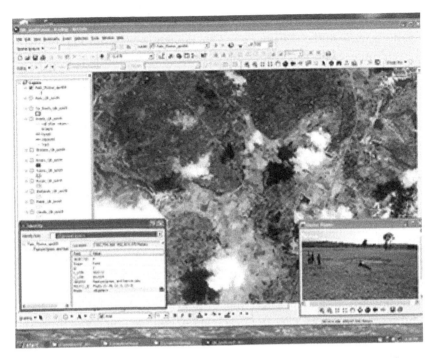

Figure 5.7. Satellite image of Waktola: Ground truth. (GIS image courtesy of Magaly Koch.)

factor in malaria transmission.[27] The lab team back in Boston measured the prevalence of malaria infection in the mosquitoes by looking closely at the infection via the DNA of *P. falciparum* and *P. vivax*. The team of Ethiopian project staff set up in the local clinic that carried out larval and pupal surveys at weekly intervals at each site by means of standardized dipping and rather clever new netting techniques. This was the work and insightful observation of experienced staff like Wolde Mikael, who had trod those rural paths for years.

We anticipated that in an area of unstable malaria transmission, the potential effects of maize on vectorial capacity might sometimes hide fluctuations of the infection in the human bloodstream. How many local folks who showed no symptoms actually carried malaria parasites? We sought, instead, to assess the potential influence of maize on mosquitoes' threat by measuring physical characteristics of the vectors' wing lengths. Better larva nutrition increases the size of the resulting adult mosquitoes, and the larger and more robust mosquitoes may then have increased daily survival rates and relative fecundity.[28] Although the study site in southwestern Ethiopia is relatively close to the equator ($7°53'$ N), the temperature during the main transmission season (August–October) tends to be surprisingly cool (averaging about 18 degrees Celsius) as a function of the elevation (near eighteen hundred meters). Such cool temperatures slow the development of the mosquitoes and dramatically extend the interval from their taking human blood to infectiousness. Indeed, at the average temperature inWaktola, the extrinsic incubation period of the parasite's period of growth approaches or exceeds thirty days, a life span that relatively few mosquitoes can reach. Maize pollen would produce more vigorous larvae that would in turn develop into larger, more robust adult mosquitoes. Our study expected that the local force of malaria transmission would help the maize pollen–fed adults.

To understand maize pollen's role in setting the size, life span, and survival of An. arabiensis mosquitoes, the project team: (1) measured wing lengths of mosquitoes collected as pupae captured from habitats near and distant from fields of pollen-shedding maize; (2) dissected female mosquitoes collected from homes near and far from pollen-shedding maize to investigate daily survival rates from each group and the number of times she produced eggs; and (3) sampled larvae from sites near and distant to fields of pollen-shedding maize, and the survival of each when deprived of food sources, like pollen. In the end this provided useful information on the potential contribution of maize pollen to mosquitoes' malarial power.

What did we learn from this place over five years? Our preliminary analyses of samples collected during the first and second seasons (during which we reversed pollen control) were most encouraging and often compelling. Traps placed within homes captured more than ten thousand adult mosquitoes during the first work season. The malaria mosquito that most worried us, Anopheles arabiensis, made up more than 80 percent of the total sample. In some homes, the traps captured more than three hundred An. arabiensis in a single night, which confirmed that this mosquito was impressively abundant and that biting pressure from this malaria vector was particularly intense in this place. We also noted considerable variability in the abundance of this vector between homes within study areas, and even within individual homes from day to day. Although the data during the first years of effort showed a trend toward more mosquitoes on tasseled, pollen-bearing sites, this apparent trend changed during the latter years of study. The study results were a bit confusing. In some years, adult An. arabiensis were abundant within homes near tasseled maize fields as well as homes near detasseled fields. There was substantial variation from year to year in many of our measurements—another sign of unstable malaria.

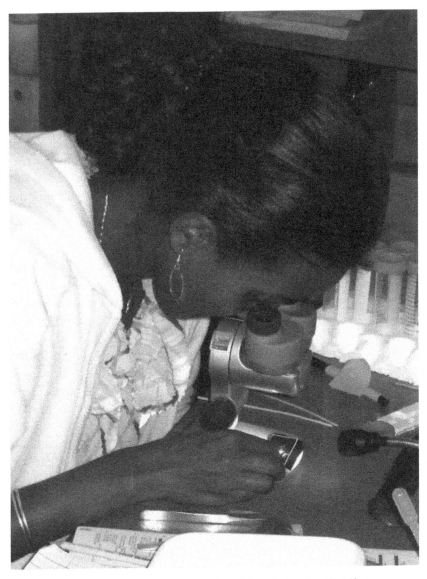

Figure 5.8. Laboratory measurements in Asandabo of wing length. (Photo courtesy of Richard Pollack.)

Another factor of confusion is that mosquitoes can fly up to seven kilometers—more than five miles. But they don't do so if they don't have to—they are efficient energy users—and maize pollen available nearby makes such travel unnecessary.

Advice from maize breeders' studies of maize pollen dispersal and conversations with specialists from the International Centre for the Study of Wheat and Maize in Addis Ababa and Mexico City resulted in the agricultural test we used.[29] We had anticipated that our intervention to control for pollen's effects would measurably reduce the abundance of adult *Anopheles* within houses on intervention sites. Although we could not detect a consistent reduction in numbers of mosquitoes, our basic hypothesis about a link between maize pollen and malaria transmission risk seemed a likely result, even if annual fluctuation matched the unstable annual appearance of malaria itself in this highland setting. Maize pollen fuels mosquito population growth. And we learned a great deal about the role of maize pollen in the local ecology of malaria.

Anopheles Adult Size and Numbers

Project staff working in a makeshift lab in the local clinic at Asandabo spent long days at microscopes, measuring the wing length of adult female *An. arabiensis* sampled from homes in pollen-reduced sites. We also captured adult mosquitoes from homes in areas of normal cultivation in pollen-rich sites, and also those we raised in the lab from pupae sampled from puddles within those different sites. The team set up lab protocols for collecting wing measurements. During the first years of work, we noted that wings of mosquitoes collected from many pollen-reduced sites tended to be smaller than those from homes in normal pollinated sites.[30] Because longevity is the most significant factor for the mosquitoes' probability of transmitting malaria, it is likely that these smaller mosquitoes would contribute less to malaria transmission than their counterparts sampled in the fully pollinated sites. This result appeared to support the link between intensive maize cultivation and malaria transmission.

The team therefore, yet again, had to deal with what unstable Ethiopian malaria looks like up close. The size of the adult *An. arabiensis* was not markedly different, whether the house was on a tasseled site or a detasseled site. Perplexing, but was this part of the unstable natural of malaria ecology in these places? Although we did not expect this result, it encouraged us to investigate other possible ecological explanations for the spatial and temporal associations between maize cultivation and malaria risk in our sites.

Ethiopia's microecologies, as the Italians had learned, don't conform to simple patterns. In Waktola, project researchers monitored the abundance of anopheline larvae twice each week in selected habitat sites. Overall, *Anopheles* larvae were far less abundant on detasseled sites than on control (pollen-rich) sites where we allowed the maize crop to shed pollen freely.

Whereas a seasonal peak of larva numbers was apparent in the comparison sites, anophelines were also present at reduced and fairly steady levels at the pollen-rich sites. These data did not fit our prediction. Why? A puzzlement. The early laboratory work and field statistics would have predicted that the reducing of pollen amounts, by starving larvae, would cause them to survive less often rather than develop to pupae and then to emerge as adults. We needed to try other studies of the local ecology to understand the nutritional value of other edible stuff within habitats where maize pollen was abundant and where it was not. This is the complex dance that played itself out in villages and households in ways that laboratory results cannot fully anticipate.

Anopheles Larval Starvation

Larvae's resistance to starvation conditions—like reduced pollen—should give an idea of the kind and amount of food available to mosquito larvae in puddles, hoofprints, and borrow pits across the farmed landscape. With this in mind, the team then studied larvae's development in maize pollen–rich habitats where they found better nourishment than their counterparts in pollen-reduced habitats. Toward this end, the project collected larvae from natural developmental sites throughout Waktola during August–September 2010. On a daily basis we studied the viability of larvae, and field staff in the local lab recorded the number of days each survived without feeding.[31] Results? In the end we concluded that larvae from maize pollen–rich sites are better able to survive and develop than their agemates in pollen-deprived puddles. This is a small step in understanding the full dynamic of mosquito life cycles in a world of abundant maize. But there is more. . . .

Nutrient Analysis of Maize Pollen

Our project staff had previously reported from controlled laboratory observations that maize pollen is a spectacularly nutritious food source for mosquito larvae. But it may be that maize pollen not only provides nutrition directly to mosquito larvae, but also indirectly by nourishing microbiota within puddles and hoofprints that mosquito larvae then eat. Originally, we had not thought about that, nor had the literature on mosquito entomology and agriculture. Science research sometimes favored simple, single-strand connections rather than ecological complexity. So we sought to analyze the nutrients and dynamics of microbial degradation of the pollen nutrient source. It became obvious. Evaluating the food quality of mosquito larvae and other parts of their food chain was an important addition to understanding the dance.[32] And here was the main point for us: The dissolved contents of maize pollen would be rapidly and easily available to nourish mosquito larvae and microbes within the environment.

Team member Bezawit Eshetu's analysis of pollen as food surprisingly demonstrated that rather than pollen merely being a DNA vessel, it is high in bulk (150 kilograms per hectare), and that it is also a highly nutritious food source containing 19 percent protein, 68 percent carbohydrate, 2 percent fat, 2 percent ash, and 9 percent moisture. A time-series analysis for release of water-soluble contents confirmed that virtually all of the soluble contents diffuse out of the pollen grain within about two hours of contact with water.

Bezawit's diligent study of the mosquito larvae's watery world confirmed that maize pollen does, indeed, serve as a food source for microorganisms. Maize pollen fully dissolved in water via activity of various microbes in about three days. Seeing this led us to suspect that abundant quantities of maize pollen in Anopheles developmental habitats can nourish mosquito larvae directly, but also support dense populations of other microbes that may, in turn, further feed mosquito larvae. Bezawit's meticulous observations won her an MSc award in microbiology, but also showed us the role of maize pollen as a direct and indirect food source for anopheline larvae.[33] Maize pollen was not the cause of malaria, but it certainly was an enabling partner!

Here is the issue: Mosquito larvae consume diverse particulate matter trapped by the surface film and suspended within the water column of their aquatic habitats. Some proportion of the food would include microorganisms such as algae, bacteria, and fungi. We sought to categorize and describe the predominant microbial flora within anopheline developmental habitats of our study sites. Team entomologists selected habitats that differed in the turbidity (clearness) of water and exposure to sunlight, and we sampled water to see its microbial diversity and density. The team then went one step further to understand more deeply the why and wherefore of breeding sites.

Borrow Pits and Malaria: A Life History[34]

What are the farm subecologies that affect malaria transmission? Borrow pits are among the best breeding sites for Anopheles arabiensis. They are small, shallow pits of perhaps a meter or two in diameter that women, and sometimes men, dig to extract special clay soil used in making pottery or making house wall plaster. These sites also serve as habitats for other kinds of mosquitoes and for diverse insects, crustaceans, and amphibians. The pits themselves vary in size, shading, depth, and the extent of vegetation supported.

Each pit undergoes remarkable changes between years and also within any given season. When newly excavated, a pit generally has steep sides and is free of emerging vegetation. Under daily conditions of rain, flooding, and animals tromping over the sites, the vertical sides (produced by use of spades and agricultural hoes during excavation) begin to sag. During one or more seasons, the pits progressively become shallow as they collect silt, increasingly support vegetation, and become low in oxygen (eutrophic) but rich in algae and even tadpoles and water bugs. And not all borrow pits are equal: Pits support anophelines, but the progressive changes in the contents of each pit seem to fall into intervals during which a pit supports different creatures, including anophelines and other mosquito types. Entomologists in our group wanted to understand the physical and biotic features that indicate viable anopheline habitat. Accordingly, we described the physical features of an array of pits and sampled plants and animals from each at different times of the season. Walking through farms daily at that very local scale gave clues to unappreciated ecological relationships in the microecology of value to local vector control and agricultural development.

The project team surveyed an array of eighty-two borrow pits within our study sites. We wanted to see ecological changes and overlapping biological successions within the pits that affect macroinvertebrate communities and, in particular, the abundance of the *Anopheles arabiensis* mosquitoes that developed as a part of that mix. In 2011 the team site mapped, categorized, and sampled: (1) during the middle of rainy season (August); (2) during the transition between rainy and dry seasons (September); and (3) at the beginning of the dry season (October). Rich Pollack and the field team then sampled sites by standard means using traditional mosquito dippers as well as by the novel means of mesh nets.

The borrow pit micro-study turned up fascinating information on the secret life of malarial mosquitoes. *Anopheles arabiensis* and *An. coustani* were the sole malaria vectors we found in these sites, and they often coexisted with *Culex* spp.—the type that buzz around your ears at night. Among the eighty-two borrow pits sampled in August 2011, *An. arabiensis* larvae were found in 40 percent, while *An. coustani* were found in 39 percent. Both kinds of anophelines coexisted in 17 percent of the pits.[35] *An. arabiensis* were more abundant in recently excavated pits, whereas *An. coustani* were more abundant in pits that had shown clear signs of erosion, siltation, and vegetative growth. *An. arabiensis* is the main vector of malaria in this community and mainly found with recently excavated pits. As the puddles age, *An. coustani* predominates, and, significantly, these are less important as malaria vectors. The ecological age of Waktola borrow pits, therefore, has a strong effect on the abundance of malaria vectors, and so, on the local force, of

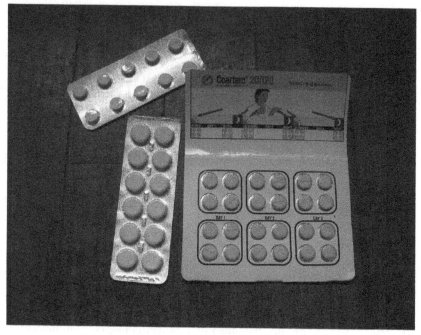

Figure 5.9. Antimalaria drug therapy (2009). (Photo by author.)

malaria transmission. Older pits were less dangerous than new ones? Borrow pits themselves, it seems, were an active, changing participant in the dance.

Anopheles Food and C4

Some more lab science. The maize-malaria link came down to tracing the genetic footprint of the maize plant itself. Maize pollen may feed the new generation of mosquitoes as youthful larvae, but it may also mark them for life.[36] We can distinguish the plants and their products (including pollen) as C3 or C4 by analyzing samples for their signature carbon isotope profiles. Animals incorporate into their own tissues a carbon signature consistent with that of the plant source upon which they feed. Mosquito larvae that feed upon maize pollen thereby incorporate such markers and retain these signatures even after they've matured to the adult stage: This offers a really useful tool for evaluating how much maize pollen nourished .

The team sampled larvae from each habitat before, during, and many weeks after the peak pollen shedding time. At each time point, we also sampled the suspension of particulates within the larval habitat and adults within homes nearby, and we analyzed those dried samples for their 13C isotope profile. The goal was to see maize pollen's role within this complex system.

Interesting. Our analyses confirmed a greater contribution of C4 to the larval diets of adult mosquitoes sampled near maize fields, relative to those sampled at distant sites. Most noteworthy was that a significant C4 signature (i.e., maize) was detected in many mosquitoes sampled from within homes. We detected an increase in the C4 signature in larval mosquitoes sampled during and shortly after the peak maize pollen–shedding period. This seemingly obscure fact adds further weight to the conclusion that maize pollen provides significant nutriment to mosquito larvae.

More on Method: Mosquito Spatial Detection by Satellite Image?

Walking in muddied fields, visiting farms, and talking to team members at each site told us a great deal about things local. Our ground-based observations of mosquito habitats and maize fields also helped us to explore the broader effects of landscapes and land use on malaria risk. We wanted to use satellite images to visualize this larger picture. Toward this end, we worked with geographer Magaly Koch, from Boston University's Center for Remote Sensing, to obtain a series of satellite images, at resolutions equivalent to aerial photographs, to record small-scale features such as individual houses, trees, field boundaries, roads, or tracts and even larva habitats scattered around them. We digitally "extracted" features of interest for geospatial insights. Her brilliant breakthrough for us was to put older maps and photographs into digital formats to show change from pre-satellite days to images from the week before we walked on the site. We could then see spatial associations between larval habitat, homes, roads, and fields of crops at key moments of malaria activity. We also pursued a time series of moderate spatial-resolution Landsat satellite images (TM/ETM+) spanning two decades (1984–2003), and more recent high-resolution QuickBird images (2006–2009), to identify and analyze land surface changes in the region's watershed, which includes the Waktola region.

Changing Ecologies

Another objective of our analysis was to look at land cover/use changes in the Waktola area in relation to potential malaria habitats. Looking at the satellite images allowed us to qualify and quantify the dramatic reductions in vegetated areas that occurred alongside the reservoir construction and filling, and also coinciding with the expansion of cultivated land into wetland areas—complexity of the dance. When coupled with our field surveys, these

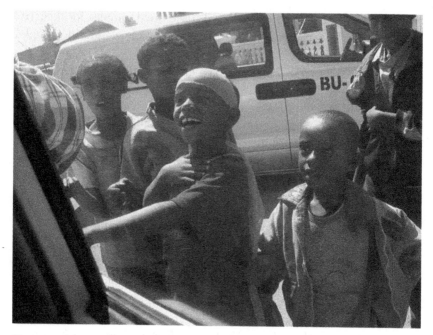

Figure 5.10. Boy with bednet headscarf. (Photo by author.)

analyses provided a powerful tool to see landscape status and transformations across diverse scales.[37] Did our choice of a local setting for studying
malaria reveal other potential insights? Maybe.

Antimosquito, Antimalaria in the Modern World

Until recently, antimalaria strategies in these communities have relied upon
indoor residual spraying (IRS) of DDT, treatment of clinical cases with chloroquine or other drugs, and some limited use of insecticide-impregnated
bednets (ITNs).

First, in 2009 a colleague at Jimma University and his Belgian collabora-
tors published a report that DDT used in the study site was effective only 1
percent of the time. What had been the post–World War II breakthrough in
the global struggle against malaria had now become useless. What's more, another study of an antimalarial strategy — insecticide-impregnated bednets —
by colleague Michal Reddy on Bioko Island in Equitorial Guinea, confirmed
our own internal study of our Waktola farmers, that bednet use was very low
(31 percent in our case) and declining.[38] These antimalaria strategies did not
work in the real world of rural folk.

Figure 5.11. Bednet use (as ox tether). (Photo by author.)

Along the way, we sought some local solutions or tested alternative antimalarial strategies. Colleagues based at the University of Vermont but
working at the Waktola site began an investigation to test locally derived
botanical insect repellants. The chinaberry (*Melia azederach*) is a tree that
occurs in areas of Asia and Africa and exists in many malaria-endemic zones,

including the Ethiopian highlands. African farmers often plant chinaberry trees because they are a fast-growing good source of firewood, and people welcome the shade they provide in arid landscapes. Extracts of the tree and its seeds repel mosquitoes and other insects. Maybe this will be a local contribution to malaria control among other new avenues.

At our site in Waktola the team collected local chinaberry seeds, dried them, and pulverized them to a powder. The Vermont team then tested the powder in bioassays in our laboratory in Asendabo utilizing locally collected larvae of *Anopheles arabiensis*. Chinaberry seed powder inhibited the emergence of more than 93 percent of mosquitoes using techniques and materials farmers might find locally. These results were encouraging and need further evaluation for efficacy, practicality, and environmental impacts.[39]

Conclusion: The Agroecology of Unstable Malaria

Rockefeller Foundation support for this 2005–2011 study began with the goal of "proving wrong" the connection between malaria and maize cultivation. It had begun with the idea of testing a single relationship—the connection between intensive maize cultivation and malaria transmission. The five-year study confronted ecological complexity but found no evidence to disprove the project's initial hypothesis on maize pollen's effect on larval survival and its role in building vectorial capacity. Our own study took place within what we recognized as a quite complex set of moving parts. A dance floor in motion.

Maize does not cause malaria, but it can and does intensify its spatial distribution and the periodic intensity of epidemic human suffering. That study of malarial landscapes at the local level also revealed an astonishing complexity of ecological elements.

The historical record demonstrates that periods of malaria quiescence and epidemic outbreaks are persistent patterns, if not predictable, and are affected by rapid agroecological change. New elements of ecological change at work in Ethiopia's dynamic human and economic setting can and will affect malaria's future. Elements of change included new land uses (such as new crop ecologies); population movements (i.e., human reservoirs of parasites moving around); and vector/parasite resistance to spray and drugs. Malaria thrives during ecological disturbance of economic growth, political instability, and interruptions of control efforts. Investments in vector control, public education, and monitoring of vector habitat will continue to be essential (see epilogue).

The dance has now taken on new actors. One subtle but intriguing change is the shift in mosquito species roaming the areas. In Waktola the collections in the past few seasons began to show *An. coustani* in addition to *An. arabiensis* in puddles and pits. This change is to a species considerably less dangerous as a mosquito vector than five years earlier when *An. arabiensis* dominated our collections. In Bahir Dar, the northern study site, the opposite might be true, as at the lakeside the longtime residents *An. pharoensis* and *An. funestus* may have lost ground to the more dangerous *An. arabiensis*, which thrives in the waterlogged construction sites of the emerging city's growing urban footprint. There was an outbreak of vivax malaria in the town in January 2010, and in February 2013 cases persisted in spots around the city as I visited local clinics with my partner Asnakew Kebede. Yet the urban setting meant that nearby clinic staff could diagnose and treat malaria fevers quickly and with the right medications. This is progress, but nothing that promises eradication (see afterword).

CHAPTER 6

She Sings

A Mosquito's-Eye View of Malaria

She doesn't buzz, she sings.

—Ato Sewnet Tegegne, retired malaria worker, Burie (2005)

A. gambiae has been found in 13 of the 14 provinces that comprise the Empire. The ubiquitous nature of this species is evidenced by the fact that it has been encountered almost everywhere that it has been sought. Its preference for small, natural unshaded collections of water, such as produced by rain, permits both extensive and intensive development. Temporary breeding places created by man are generally recognized as being among the most important to the species.

—Charles O'Connor, WHO malaria specialist (1967)

This species [A. gambiae Species B] is the most widespread, and individual females tend to be either endophilic or exophilic, anthropophagic or zoophagic, early biters or late biters, and doubtless other alternatives, according to the arrangement of their floating chromosome inversions. . . . Genetic polymorphism of Species B may lead to true behavioristic resistance.

—G. B. White, Wellcome Parasitology Unit, Addis Ababa (1974)

MOSQUITOES ADAPT AND ARE the soul of malaria. They move, they experiment; some die as they try new strategies to survive a sprayed chemical (like DDT); others succeed, passing on their genetic behavioral successes to a new generation. A few individuals deviate, but entire species or subspecies emerge and adjust to a local climate's dryness or new blood sources. She may bite indoors or out, at dusk or 2:00 a.m. when her blood-meal victim is fast asleep. She usually lives just a few days or, rarely, a few months. And she may fly seven kilometers to a meal or only a few meters, and over the course of a human victim's life that person may donate blood a few times or many thousands of times. She may take a blood meal from an animal or a human. And in these adaptations there is something of a collective spirit, but one where individuals pursue their own needs.

The story now pulls malaria away from the strained voices of optimism of the World Health Organization and international agencies like Roll Back Malaria, the Global Fund, and USAID into the most local perspective possible—the testimony that might be offered by the mosquito herself. If asked, the mosquito herself might say, in a small voice:

> There are many types of us: We dance, dodge, and dart in different ways. In Africa there are only a few of us that matter, those who can carry malaria. About forty-two cousins and subspecies, but really only about five types of us that transmit malaria to humans with any efficiency. And sorry, but it's unintended. Humans give us the parasites, by the way. We only pass them along. Those parasites burden us, too, but only mildly.

Other voices in that local malaria chorus might include the parasite, and the human patients who have responded in their own ways to new ecologies of environmental and political change. Here the story shifts between the imagined voice of a female anopheline mosquito who might celebrate her own brilliance at adaptation, even while she scoffs at the persistent but premature claims about malaria eradication from human health-care bureaucracies. She, and others like her, live, reproduce, and evade. She is, in effect, a wild animal, like the suburban possum, which has coevolved with her human fellow travelers and the changing ecologies of the ancient and modern worlds. Her survival has made her "plastic"—nimble, clever, and resourceful.[1]

The mosquito is an ironic model of both individualism and a belief of safety in numbers. And a responder to human behavior. A view from entomology helps here, since that view of nature divides creatures—including mosquitoes—into an overall set of types, depending on their survival breeding strategies and the role of natural selection. In the language of ecology, the r-strategy describes mosquitoes, since they succeed in survival and reproduction by large numbers and selection pressures that allow them to adapt by quick evolution to changing conditions. The k-strategy is that followed by mammals, some birds, some insect vectors, and . . . humans:

> R-selected species have a population size that is limited by repro-
> ductive rate (r), density independent, and relatively unstable. The
> organisms in their population are generally smaller and short lived,
> and have a relatively early reproductive age, produce many off-
> spring, and provide no care for their offspring. Examples would be
> locusts, flies, bacteria, algae [and mosquitoes].
>
> K-selected species have a population that is limited by carrying
> capacity (k), density dependent, and relatively stable. The organ-
> isms in their population are larger and longer lived, have a relatively
> later reproductive age, produce fewer offspring, and provide greater
> care for their offspring. An example would be elephants, wolves, oak
> trees [and humans].[2]

Mosquitoes survive and adapt by massive egg production and inattention to
parental guidance for the young. The female mosquito not only perpetu-
ates the species by passing on her behavioral genes, but her own role in
malaria transmission exerts selection pressure on humans whose bodies and
choices of where to live adjust to that disease's effects. If the mosquito had
a voice, we could listen to her as she chided her hatching young ones. She
might have a few hundred hatchlings that had survived their early life in
the puddle or hoofprint, and she could explain to them in simple terms that
they were an "r-selected" species. By laying hundreds of eggs on the surface
of a watery habitat, she expected that some would find food, dodge predators, and survive to take flight out of the puddle in the few weeks of life, to
seek adult food in the form of nectar and then, if female, to find a one-time
male partner to fertilize her eggs before she took off on her own to repeat the
cycle. Then she might advise the young ones to find blood from a human,
or maybe from a calf or a sheep. Filling her belly with blood and finding a
place to rest, she would excrete the blood's liquid and leave the proteins to
nourish the eggs until she was ready to take flight again, then find a puddle
or hoofprint in which to lay her eggs and wish them luck—the foundation
for a new generation. Just a precious few of her young need to survive to as-
sure that her lineage continues.

But the mosquito mother might also want to explain, honestly, that as an
r-selected species there is no need for parental advice to the young nor for
much care in selecting egg-laying sites beyond, perhaps, presence of water
and sunlight/shade conditions. She doesn't much calculate survival of her
young ones. Mosquitoes succeed not by education and parental protection—like elephants, musk oxen, or mammals—but by the sheer force of
numbers and collective instincts assembled over generations of selection.[3]

What matters most to the young mosquito's survival is not parental nurture, but the group's genetically driven behavioral makeup in the species individual's amazingly quick adaptation—in a couple of new generations—by employing new strategies about when and where to leave her eggs, whether to seek her blood meal indoors or out of doors, when to seek that blood, and where to rest and excrete her unnecessary liquid, leaving the blood cells to nourish her developing eggs.

Above all, for malaria transmission, it matters whether she takes her blood meal from a human, an animal, or either. She can get nourishment for her eggs from any warm-blooded creature, but gets the malaria parasite only from humans. With all of these prerequisites, one sometimes wonders how malaria persists so remarkably well.

Anopheles Species B (i.e., *An. arabiensis*) takes blood from humans and other mammals. This female's genetic makeup predisposes her to lay eggs in large numbers on watery surfaces and let them fend for themselves. The larvae that hatch instinctively hide from the shadows of predators (see chapter 5 on Waktola) and seek nourishment in the forms of algae and bacteria—or perhaps nutritious pollen grains from nearby fields of maize. Mosquito larvae are, after all, often food-limited rather than being diminished by predators that feed on them. There is provocative evidence that mosquitoes that seek their blood meals from nonhuman animals (zoophilic) have a lower survival rate than those that hunt human blood.[4] Do animals' tails swat better than human hands? Perhaps. Or do other mammals perhaps sleep more lightly? Future research must tell us that.

After all, slaking those bloodthirsts is her prime directive, not transmitting malaria parasites. But the when, where, and what kinds of blood and how the young larvae survive depend very much on factors like moisture from rainfall or water sources, and temperature, and the presence of a dark, cool place where she may rest while her eggs mature in her body. Generations of selection and being swatted have encouraged those behaviors. And rural ecologies offer a wide variety of land uses, crop types, water havens, and potential sources of food for the young ones. Unlike K-selection beings, the female mosquito does not seem to be very selective. Her not-so-subtle philosophy: Lay eggs everywhere and hope for the best (though we have no evidence that egg-bearing female mosquitoes actually express "hope").

At the turn of the twentieth century, scientist Ronald Ross pointed out in his monumental 1911 work, The Prevention of Malaria (with Addendum on the Theory of Happenings), that the transmission of malaria is extremely sensitive to the survival rate of the vector.[5] In truth, though mosquitoes behave, survive, reproduce, and adapt as a collectivity in some ways, they act solely as individuals in another. Social insects (called "eusocial" by entomologists) like ants or honeybees act in genetically defined "caste roles" around female leadership, functional niches, and a common purpose of collective survival.[6]

Edward O. Wilson implicitly draws a sharp distinction between mosquitoes and ants. He is a fan of ants. For ants' social instincts, he argues, "Karl Marx was right, socialism works, it is just that he had the wrong species." Wilson's meaning is that while ants and other eusocial species appear to live in communist-like societies, they do so only because they are forced to by their basic biology, as they lack reproductive independence: Worker ants, being sterile, need their ant-queen to survive as a colony and a species, and individual ants cannot reproduce without a queen and are thus forced to live in centralized societies. Humans, however, do possess reproductive independence so they can give birth to offspring without the need of a "queen," and in fact humans enjoy their maximum level of Darwinian fitness only when they look after themselves and their offspring, while finding innovative ways to use the societies they live in for their own benefit.[7]

Ants as a eusocial "society" have another characteristic that likens them more to humans than to mosquitoes: Their societies have assigned (by genetics) roles in the social hierarchy: soldiers, workers, a queen, and—as Edward O. Wilson and Bert Hölldobler call these social divisions—subcastes. Wilson and Hölldobler's study of leafcutter ants (Atta) identified skills of social communication between types of ants in a colony that amount to a bureaucratic structure.

One of the most obvious differences between the eusocial ant colonies and mosquito populations is what Hölldobler and Wilson describe as "mutualists," where ants cultivate fungus in their tunnel networks as food and in turn feed on that fungus, using techniques that resemble agriculture, and something chemical that approximates group communication. In one fascinating experiment they describe, a poisonous element (an orange peel treated with a fungicide) was offered to foraging Atta workers/collectors as a potential food for the colony. Foraging workers then brought bits of it back to the main anthill where its poison harmed the resident fungus. Somehow reading the fungus's distress at the poison, the foraging ants from the colony then passed a chemical message back to the collectors to reject the orange peel as a viable food. They eventually got the message, but the rejection behavior took ten hours to begin after the first poisons entered the hill. The rejection continued for nine weeks, even after the poisoned orange peels stopped arriving. The communication system had had a delayed start and only ended after a longer period. This "bureaucratic" reaction/response actually took place over several intercaste communications and over several ant generations. The delay in recognition, policy adoption ("no more orange peel"), and policy change (a "coast is clear" message once the peel was made fungicide-free) was a social communication event quite unlike the response from mosquitoes who would have responded not by communication but by selection pressure. In other words, nonadaptive mosquitoes would die and only those with the adaptive "skills" would reproduce in future generations.[8]

Mosquitoes, by contrast, act as what might seem like intuitive individuals from their hatching as larvae to their procreative adult lives. And it is their sheer numbers seeking to live, survive, and procreate as individuals with genetic diversity. Malaria the disease has coevolved with mosquitoes being the vector. In the century that followed Ronald Ross's pioneering ideas about mosquitoes from his work in the Indian colonial service, there had been slow progress in human science, but steady, clever movement by mosquitoes themselves in Ethiopia and elsewhere. The chess game played by humans against malaria consisted in large measure of the suppression of mosquitoes by spraying toxins and also by sorting out exactly what kind of mosquitoes they were according to their body types. The mosquitoes, however, played their game of swirling survival, attack, and defense (retreat).

Ethiopia's landscapes were a part of that effort as an object and participant in the play. But again, science and human health bureaucracies played chess, and the mosquitoes were in an elaborate, fluid dance of adaptation. Perhaps ants as eusocial communities require time to learn and respond to the danger of a poisoned orange peel. That response took ten hours to stop the collection and three weeks to return to collection of the nonpoisoned peel once the poison disappeared.[9] Human collective responses to learning about mosquitoes, malaria, and their ecology took place over a long period of time and within bureaucratic structures like a Ministry of Health, the World Health Organization, and research laboratories. That bureaucratic response took eight years once they learned about the parasite's resistance to the chloroquine drug.

Changes of behavior belie the misleading chess metaphor. The evolutionary movements and behavioral changes among anopheline mosquitoes were on several levels: zoophily versus anthropophily (blood meals taken from animals or humans); exophagy versus endophagy (whether to bite outdoors or indoors); endophily versus exophily (whether to rest indoors or outdoors after the blood meal); and the ability to tolerate dry conditions and then spring into life quickly once rains began. If the malaria problem was a chess match, then it was a *Star Trek* version of the game with multiple levels and dimensions—of space and of time. And human understanding of these shape-shifting behaviors only had its early stages in the first half of the twentieth century.

Overall, the patterns emerged in competition between larvae in the water pools and in adult preferences for blood meals. Human scientists and field teams struggled across the globe and in Ethiopia to identify and relate what mosquitoes looked like and the way they behaved. They especially sought to understand which ones carried malaria and how likely their bites were to inject the deadly plasmodia (*P. falciparum*) that dominated in Africa or the sickening one (*P. vivax*) that appeared in a place like northern Europe where conditions were cooler.

The first identifications of the anopheline species in Ethiopia actually predate the Italian invasion and the earliest recognition of the anopheline mosquito–malaria connection in the first decade of the twentieth century, and continues through the Italian colonization of 1936–1941. The list of thirty-five species and two subspecies is long: The names and the dates of first collection tell a bit of the history of Africa and the Nile Valley as a malarial zone.[10]

More importantly, this cacophony of mosquito types really masks the fact that only four of them played a significant role in malaria—and the main one does not yet appear on this list. That monster—*An. arabiensis*—had made its (or her) public appearance just the year before and in a laboratory at London's Keppler and Malet Streets (London School of Hygiene and Tropical Medicine), a few hundred meters from the Rosetta Stone in the British Museum. Was her identification a similar breakthrough?

Actually, *An. arabiensis* had been in Ethiopia and much of East Africa all along. Dr. George Davidson, an entomologist working at the London school, had been maintaining a colony of what seemed to be *An. gambiae*, Africa's most deadly and ubiquitous mosquito. Using classic lab techniques for experimental breeding, he had found that some males were infertile when he crossbred specimens in the lab to see if they were actually distinct species that only looked similar under the microscope but were genetically distinct. In the December 1962 issue of *Nature*, he published a rather timid idea in a single-page note, saying that "it would appear that Anopheles gambiae is a complex of at least four partially incompatible forms. Whether any or all of these deserve rank as separate species remains to be seen."[11] The tone of the piece seems polite, almost apologetic, yet it opened the game to some new rules that the mosquitoes seemed to follow.

Species (reported by, year identified)	Species (reported by, year identified)
A. *adenensis* (Christophers, 1924)	A. *macmahoni* (Evans, 1936)
A. *ardensis* (Theobald, 1905)	A. *marshalli* (Theobald, 1903)
A. *christyi* (Newsted and Carter, 1911)	A. *natalensis* (Hill and Hayden, 1907)
A. *cinereus* (Theobald, 1901)	A. *nili* (Theobald, 1904)
A *coustani coustani* (Laveran, 1900)	A. *obscurus* (Grunberg, 1905)
A. *coustani tenebrosus* (Donitz, 1902)	A. *paludis* (Theobald, 1900)
A. *coustani ziemanni* (Grunberg, 1902)	A. *pharoensis* (Theobald, 1901)
A. *danacalicus* (Corradetti 1939)	A. *pretoriensis* (Theobald, 1903)
A. *demeilloni* (Evans, 1933)	A. *rhodeniensis* (Theobald, 1901)
A. *d'thali* (Patton, 1905)	A. *rivulorum* (Leeson, 1935)
A. *funestus* (Giles, 1902)	A. *rufipes* (Gough, 1910)
A. *gambiae* (Giles, 1902)	A. *rupicolus* (Lewis, 1937)
A. *garnhami* (Edwards, 1930)	A. *seydeli* (Edwards, 1912)
A. *harperi* (Evans, 1902)	A. *squamosus* (Theobald, 1901)
A. *implexus* (Theobald, 1903)	A. *theileri* (Edwards, 1912)
A. *kingi* (Christophers, 1923)	A. *turkhudi* (Liston, 1901)
A. *leesoni* (Evans, 1931)	A. *wellcomei* (Theobald, 1904)
A. *longipalpis* (Theobald, 1903)	

The real splash actually came two years later. Davidson's article in the *Bulletin of the World Health Organization* shook the foundations of Africa's malaria map, though strangely it took awhile for it to penetrate the bureaucratic human brain tissue of Geneva, Washington, and Addis Ababa. Davidson's article's modest abstract read, in part:

> The author reports on some 200 laboratory crossings of 36 strains of *Anopheles gambiae* from many different parts of Africa which show the existence of five mating-types in what was recently considered a single species. Three of these mating types are freshwater forms and have been provisionally called A, B, and C. . . . From a practical point of view it may be necessary for the field worker to be able to identify the exact species with which he is dealing before the most efficient means of controlling it can be found.[12]

To be sure, at the very end of the nine-page article, complete with photos of atrophied male mosquito organs and tables of crossbreeding results, he notes, again with characteristic modesty: "These features of speciation in *A. gambiae* will be of undoubted interest to the evolutionist. To the entomologist and malariologist concerned with malaria eradication in Africa they may be of vital importance."[13]

In fact, Africa's wet-dry seasonality had probably fostered this remarkable revelation of species diversity: Davidson named five different species: types A, B, C, D, and F (he was seemingly not terribly imaginative). Species B, which later acquired the more exotic name *arabiensis*, was, perhaps, the cleverest of all, since her skills at surviving seasonal dry spells and thriving on animal as well as human blood gave her the ability to hunker down in the dry season and spring forward when moisture appeared. Animal blood helped her nurture her eggs. Along transition zones of the sahel or drier areas of East Africa she could coexist with *An. gambiae s.s.* (aka Species A), which had coevolved with humans and took their blood in preference to all other sources. In Ethiopia, however, *arabiensis* seems to have been the only *An. gambiae* in town.[14]

Bureaucracies move slowly. Indeed, it would take about ten more years before the antimalaria bureaucracies in the World Health Organization, USAID, and the Ministry of Health finally recognized her different behavior in her preference in blood-meal sources, dry season disappearance, and overall cleverness. This now demanded their attention. Here the orange peel analogy from the leafcutter ants comes to mind. Rapid generational renewal and evolutionary selection pressure of mosquitoes seems to be a more effective adaptive learning tool than human policy making.

A new game was afoot or, rather, had taken wing. After at least two generations of fieldwork to identify malarial mosquito types at play in Ethiopia, it became clear that only four species were really at work in malaria transmission, and in each case she had particular habits, places, and biting times that she preferred. There *An. arabiensis* coexisted with specialized local species like *An. funestus* (perennial water at lakeside); *An. nilus* (only in the Baro River area in the southwest); and *An. pharoensis* (swampy habitats near rivers and lakes). Only *An. arabiensis* exists in all of Ethiopia's malarial zones. Eventually, *An. gambiae* B took on the new name among malariologists, "*An. arabiensis*," though we do not know how she felt about her "outing" with a new variety and quite a distinct personality determining whom she should bite, when, and how she hid out during long dry spells. Yet, knowing that *An. gambiae* B existed was only a single piece in a complex puzzle—or, rather, knowing that there was a new element in the dance.

Davidson had found *An. arabiensis*, but her outward appearance still did not allow anyone to identify mosquitoes that otherwise looked exactly alike under a microscope. Only the behaviors in biting times and favorite blood sources offered clues, but only after the fact and after hours in a distant laboratory under a microscope.

Yet, finding her was indeed a game changer. Then a decade later, after the 1983 invention of polymerase chain reaction (PCR) and its later use to identify genetic markers of mosquito species, the human side moved another piece forward. It became clear that she had long been present throughout dry climates of eastern and southern Africa as well as along the edge of West Africa's savanna zone. And there were other interesting revelations about past malaria outbreaks. In 2008, malaria geneticists found that a mystifying 1930 malaria outbreak in Brazil had actually been the result of an invasion of An. arabiensis in a zone where malaria had been "eliminated."[15] The An. gambiae B had found her way across the Atlantic before she was even recognized by science. Clever girl.

Finding evidence of the mosquitoes' near relatives was a new wrinkle in the struggle against malaria, but identifying the new types quickly and consistently was the real breakthrough. Before PCR, recognizing mosquito types was a painstaking, laborious process carried out in the field and in the lab. Davidson knew this with what we might now call a seat-of-the-pants entomology that sorted mosquitoes by examining their body parts—morphological comparisons—and following generations of offspring to see sterile males and who mated with whom. And human science also needed to observe ecology: a hodgepodge of study of the relevant habitat, seasonality, and host preference. In the 1950s and 1960s, labs examined isoenzyme patterns by electrophoresis and then looked at slide-mounted and stained chromosomes for banding patterns that would help separate mosquitoes into distinct groups. Very demanding work that took skill, and not everyone could do it. PCR analysis back in the lab took out the guess-work, though it meant, at first, carrying mosquito carcasses in alcohol vials back to a laboratory in Boston, London, or Paris. Later in the 2000s, it was possible to take mosquito bodies smashed onto filter paper back for analysis (as I had done in August 2006).

Anopheles arabiensis (the mosquito formerly known as Species B) was now exposed as malaria's transmitter in Ethiopia, but also as far south as South Africa. But its shape-shifting in seasons and across its cousins' habitats still made it a formidable foe.

After the innovation of Davidson's work in London in 1964 and PCR techniques in California in 1983 had seemingly unmasked the genetic mysteries of *arabiensis*, she was, in fact, no less elusive as a quarry. For in Ethiopia, at least, a generation of DDT spraying and the earnest work of local antimalaria teams up to 1974 had failed at eradication or even local elimination. She still scoffed at human profligacy and laid her eggs with impunity, exploiting disturbed government politics, construction sites, and new "borrow" pits where women had "harvested" clay for housing repair. This group also included resettled, nonimmune people who had moved, sometimes forcibly, to new ecologies under hastily arranged government resettlement programs. These confusing elements of the dance still continued.

As in the past there were consecutive years of malaria quiescence, this time in the mid-1960s and early 1970s. In the change of seasons and rising temperatures, however, again she pounced. In ten days after the late June rains began, in the late 1980s, early 1990s, 1998, and 2003, she laid eggs in turbid puddles as temperatures rose. And if the hatched larvae in puddles or hoofprints or pits found food, they survived and took wing—males in search of nectar from flowers and females in search of that same nectar, but then the blood to nourish her eggs.

As an r-selection agent, she was an individual with no social obligations to her own generation. But she did have as her instinctive passion a brief search for a mate by synchronizing her wing beat with the male for the brief encounter to fertilize her eggs. Then the need to sustain her kind by laying eggs—and in massive quantities (150–300 at a time). Through the blood meal she took to sustain the eggs, however, she might ingest Plasmodium parasites that had their own story of asexual and sexual reproduction in her body and that of her blood-meal victim. The types of blood malaria she transmitted depended on the temperature and plasmodium type (17–19 degrees Celcius for falciparum or 15–17 for vivax). She might be carrying both or just one. And, of course, luck played a role, since she had to miss a swat from an ox's tail or a dozing human, or the once deadly DDT from a spray team, as she rested on a house's inside wall after her blood meal. Some now think that she might even seek rest in a dense field of maize near the house. By 2009, she did not mind if someone sprayed that house wall with DDT or with dieldrin—what had been a tried-and-true human strategy in the late 1950s. She had now come to be immune to the usual insecticides.[16] Like water off a duck's back.

She was not a social creature like an ant or a bee that fit into a defined social role, such as a queen or a drone or a worker, to build a habitat like a hive or a hill. She was a wanderer seeking her own survival and singular goal to reproduce—limited by her own ecological niche of moisture, temperature, and what blood meals were on tap. And therein lay her inadvertent role as malaria vector.[17]

An. arabiensis is that artful dodger: She retreats quickly when the rain stops and puddles dry; that might end eggs hatching for that season, but still allow older females to cavort for a few weeks or months after the rains have ended in the fall or spring, taking blood meals and passing along malaria parasites. And she is a quick starter once rains begin again and she reemerges zombie-like from seeming oblivion. In West Africa's humid zone, *An. arabiensis* does not compete well, since *An. gambiae* s.s. is a full-year resident and lives closely with humans in the moist environment. But in the wet/dry oscillation of the savanna edge—and in Ethiopia's wet/dry climate oscillation, she is champion of the duck and dodge—rope-a-dope.[18]

The Dance Continues

Eradication, Vaccine, and Malaria's Ecology of Persistence

THE TWENTY-FIRST CENTURY HAS brought massive change to the long-duration patterns of history—and to malaria's modernization. In the developing world, human ambitions to transform their physical setting in the twentieth century have escalated into twenty-first-century expectations for malaria eradication. And Ethiopia is at the cutting edge in the link between rapid economic change and a new malaria ecology now in process.

What the Ethiopian government and international interests plan for water and moving people around the Blue Nile watershed is monumental, with strong flavors of what Yale's visionary scholar James Scott would call "high modernism." Malaria is an unintended consequence of development, but may well be an unexpected part of such engineering feats. The government schemes for dams, hydroelectric power, irrigated agriculture, and hopes for producing commodities like oilseeds, sesame, petroleum, and cotton fiber are ambitious, but also have an aura of a Sophocles-like hubris when it comes to human health. Within the Nile system, Ethiopia controls the high ground, and its Grand Renaissance Dam is the biggest of the Nile spigots. Harvesting water into either large reservoirs or small ones, and changing the flow of waters using tunnels and sluices on the Nile's living arteries, invites in the complexities of nature, even if it promises more reliable urban electricity than Ethiopia and the region have ever known.[1] Malaria will be a part of that future, too.

Ethiopia's new Grand Renaissance (Hedase) Dam—Africa's largest—is the most ambitious of all those dreams. These visionary projects and the inevitable landscape changes will have transformed landscapes and waterscapes. But these projects also signal a new era of modernity and economic global forces that are changing physical landscapes and human health like nothing before them. And the coupling of human and natural physical landscapes that seek to manage water's precious gift to agriculture and electric power also has changed the geography for malaria and the human-mosquito waltz that will impact the health of the next generations. Malaria will remain and be local and more resilient than human bureaucratic responses designed to eradicate or even to control it. So what is the way forward?

This final chapter reprises the book's evidence and argument for the human engagement with malaria as a local disease of place or ecology rather than one for which the resolution rests entirely with biomedicine—drugs and vaccines. Ethiopia is a prism that refracts the shape of things to come. The historical tale's epilogue summarizes the human story implicit in malaria's persistent survival in Ethiopia, and in the tropical and subtropical world overall. This story will unfold in the coming era as cities in subtropical Africa sprout like mushrooms growing up and out, spilling onto rural hinterlands, pulsing with new expanding metabolisms that will shape malaria's future.

Nevertheless, the Gates Foundation's 2007 return call for "eradication" as a global goal continues to appear in government policy statements.[2] Gates's seemingly perpetual optimism about a silver-bullet vaccine has spurred a flawed vision that eradication is an attainable goal. Is that a helpful vision, given the history and complex ecology of malaria? How might Ethiopia's case of long-term affliction with malaria, the disease's unstable nature, and its ecological footprint on historical and current landscapes help lead us forward in the long-range battle/dance? Is a single-stranded investment in biomedicine for a vaccine or a genetically engineered mosquito the answer? Or is perhaps the wiser strategy a commitment to public health education, monitoring local outbreaks, and quick response to those inevitable mini-epidemics?[3]

Malaria's history in Ethiopia offers fascinating yet perplexing insights: new ecologies of poverty, economic differentiation (rise of a monied class), urbanization, and changing urban landscapes whose populations—in Africa anyway—ironically continue to get younger, not older.

An image that I draw on here is that of the new lakeside town of Bahir Dar. Historically, once an unsettled landscape because of "fevers," it is now a bustling city spread out along a lakeside becoming increasingly devoid of its historical papyrus edges, the accustomed habitat of the mosquito culprit

A. *pharoensis*. But a real change also took place with a rising landscape of buildings, disturbed construction site puddles, and new cement-block "villa" houses without window screens, but with open drainage that quickly becomes clogged, creating fouled ditches that invite mosquito larvae.[4]

In January 2011, an outbreak of malaria took place in Bahir Dar, a new gem of a modern city. It was *P. vivax* malaria, and the local clinics prescribed chloroquine and in some cases resorted to the more drastic therapy of intravenous quinine drips. It was an even more baffling event for Bahir Dar's dense, modern health services. The new drug (Coartem) did not work on the *vivax* parasite. So they dug out the old ones (chloroquine, then primaquine and then an IV quinine drip) as a last resort. The health clinics responded and saved lives, ironically by using "the bark" (quinine) of former times.[5]

What is next? In May 2012, there was an outbreak south and west of Bahir Dar at Chagni, and south Gondar, both lowland hot spots that had historically been places where epidemics had begun and then rapidly washed over the highlands. The infection rate was astonishing (43.5 percent had plasmodium parasites and symptoms), but so was the speed of diagnosis by workers with the Carter Center and the Ministry of Health, who arrived to push for a new surveillance method (ESS, or enhanced surveillance system) and a new diagnostic tool (rapid diagnostic test, or RDT) that connects local observations to national response and drugs that fit the type and severity of the malaria in a place and time. This site of the hot spot was the same from 1953, 1958, 1964, 1991, and 2003, and perhaps from long before that. But the newer, effective approach was accurate diagnosis and rapid response, not a universal vaccine.

While the rate of infection in this sudden 2012 epidemic in the west of the Blue Nile Basin was extraordinary in its quick spread, the death rate was low because of the new tools and rapid response.[6] Might this local detection of a local outbreak and response be the future of smart malaria policy, rather than a false hope of eradication? Could the future be an efficient and adaptive surveillance methodology that uses a network of local reporting, rapid diagnosis, mobile phones, and a growing body of local knowledge?[7]

Malaria is not only local, but also geographically elusive in its shifting ecological footprint whose dance steps move across spaces and times; now here, now there, turning in new ways. In Waktola we see a different ecology than in the Blue Nile Basin and perhaps different outcomes in rural and peri-urban settings. But a local ecology shows new mixes of the modern within the local. A new road across Waktola drained old swamps and encouraged "modern" crops. Its black wetland soils have dried out and new hybrid maize plants appear along the old swamp edges. In this case, high modernism took the form of a new asphalt road that would open national markets for new crops. Road engineers, sensibly, from their point of view, designed culverts beneath the road to drain what had been wet areas to save the road from heavy rain erosion. In the eyes of a road engineer's plan, it worked. But at least part of the surrounding malaria habitat changed, and unintended health consequences pushed in both directions. Wetland drainage allowed new prosperity through agricultural change and farmer entrepreneurship (maize monoculture) and a new road to market, but also brought changes in malaria's world. What had been the prime breeding territory of the menacing *An. arabiensis* now also seems to host what had been a relatively unknown player: a mosquito, *An. coustani*, that quite likes a grassy puddle habitat and may outcompete *An. arabiensis* in older puddles. Yet *An. coustani* is a poor carrier of malaria parasites. In fact, *An. coustani*'s puddle-centered victory may bring a change in malaria transmission—fewer parasites, fewer bites. Who knew? Might malaria incidence decrease with a new mosquito in town?

What else lies ahead? The dance continues, with new parts and new partners in the disturbed ecologies of growing cities in tropical and subtropical zones, moving parasites. Ultimately, we hope and expect that malaria fears development. The disease has already fled the modern ecologies of Europe, Brazil, North America, Sardinia, and many island settings in the Caribbean and the Pacific. Africa has always been the outlier, a home of the *An. gambiae* family and ideal ecologies for transmission that caused WHO to exclude Africa from their global plan back in 1955.[8]

Eradication in Reach?

The first global strategy for the fight against malaria took off in 1955, and it invoked the word and the idea of eradication. That program had as its model the elimination of malaria in North America, Europe, and First World settings where industrialization and urbanization improved public health overall and reduced malaria transmission close to zero. But by leaving Africa (including Ethiopia) out of that mix, the global eradication part of the story faltered and failed. In a March 1, 1958, *Ethiopian Herald* article titled "Malarial Control Conference Suggests Eradication Plan," the country's primary newspaper boasted strangely that "it has now been proven that malaria eradication is technically feasible here."[9] In that early effort and in the decades that followed, the plan was to spray the chemical pesticides DDT and dieldrin. Ethiopia's resurgent epidemics in 1953, 1958, 1964, 1985–1986, 1991,

Figure E.1. Mosquito spray (1964).

1993, 2003, and especially in 1998 showed malaria's astonishing resilience and ability to recede and then recover unexpectedly.[10]

Malaria aside, the poster child for the disease eradication concept since 1980 was smallpox, that ancient scourge of Europe and a deadly ally of the Europeans' conquest of the New World. Smallpox eradication efforts began in 1967. The last endemic case was in Somalia in 1977, and WHO declared

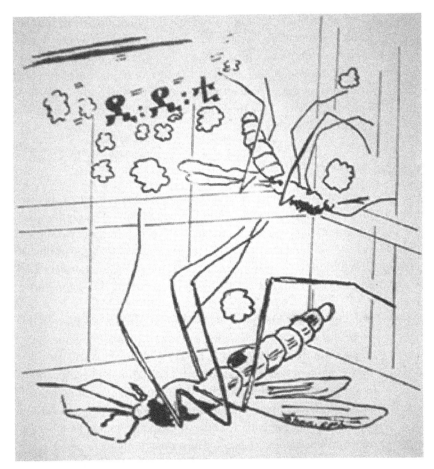

Figure E.2. A vision of mosquito death (DDT brochure, 1964).

eradication in May 1980. Was this a good model for malaria? No. Smallpox is fundamentally a different disease than malaria on biological grounds alone: Smallpox's disease-causing agent—like influenza and unlike malaria—was a virus that was airborne; its life cycle needed no animal or vector reservoir; smallpox has clear and visible signs of infection; and there was a vaccine developed to cope with the virus.[11] There is also reasonable hope for polio eradication, though the biggest obstacles are political and ideological rather than biological. Parents and political movements from Pakistan to Southern California now resist childhood vaccinations on ideological grounds. Annual influenza scares—like H1N1 (swine flu)—or bovine spongiform encephalopathy (BSE, or "mad cow disease") had their origins in human-livestock/poultry interactions, a virus-based disease ecology where disease agents moved from domestic animals like pigs, chickens, and cattle to humans by contact—a cough or a touch or in water or food. Newly emerged Ebola's virus rides on direct contact with human bodily fluids. Malaria is, after all, a protozoan borne by a mosquito vector from human to human—a different construct altogether.

Vaccine: "Sobering Results"

For Ethiopia and its unstable malaria, immunity among local people will continue to be rare and temporary. And the complexity of mosquito, parasite, and human reservoirs will keep the dance moving. In 2009, researchers at the University of Maryland's Howard Hughes Medical Institute published a lament about the prospects for a malaria vaccine. It takes us straight back to malaria's history as complex, adaptable, and coevolved. They concluded:

> The complicated biology of *Plasmodium falciparum* has presented major obstacles to the development of effective malaria vaccines. As it passes through the stages of its life cycle, the malaria parasite expresses different stage-specific antigens, each stimulating a specific immune response. Adding further complexity, *P. falciparum* has a long evolutionary history with its human host and exhibits extensive genetic diversity, particularly in the surface antigens that have been under *prolonged selective pressure* by human immune response. . . .
>
> Moreover, the parasite continues to evolve through mutation and sexual recombination in response to drugs and other malaria interventions, providing a moving target for these interventions. When malaria vaccines are deployed, "vaccine-resistant malaria" can be expected to emerge and threaten vaccine efficacy just as drug resistance has compromised the efficacy of drugs used to prevent and treat malaria.[12]

Their pessimists' conclusion used the obtuse terminology of science, "dynamics of vaccine antigen polymorphisms," to explain the faint and false hope for a vaccine solution for malaria. They may have said, more plainly, that malaria is clever in the ways that it evolves in the face of environmental change and human efforts to suppress it. But the point is the same: A vaccine for malaria has been and will be within our vision, but beyond our grasp. Therein lies the danger in the longer range.

In March 2013, the online journal *ScienceNow* published a summary of the vaccine program under the title "More Sobering Results for Malaria Vaccine," evaluating the success of the RTS,S vaccine, a witch's brew of protein fragments from the drug maker GlaxoSmithKline that went into field trials in several African countries (though not Ethiopia).[13] Ethiopia, in fact, had set a plan for field trials to test the new vaccine, and had chosen a project director, but did not carry out the plan. When no money arrived, the trial never took place. In any case, the trial vaccine required a shot and two boosters, a method that might work in a subsidized controlled trial with volunteer patients, but not in the real world of Waktola and Burie.[14]

The *ScienceNow* article also expressed disappointment that a Kenyan follow-up study with the first trial group showed that the vaccine's effectivenessn in 2011 had shown 50 percent for infants, but that even in those children the value was gone four years later.[15] Back to the drawing board. Robert Sauerein, of the Nijmegen Medical Center in the Netherlands, summed it up well, saying that RTS,S is "definitely not a big tool," but "we can't afford to throw any tool away."[16]

So where are we? We have some tools, some hope, but no panacea. Perhaps *persistence* will continue to be the watchword, rather than a false expectation of eradication. And the way forward must include valuing the vision of looking back. For Ethiopia that means accepting the unstable nature of malaria in places that include a kaleidoscope of ecologies of temperature and mosquito species that are "local." Local knowledge understood that as early as the eigtheenth century. Learning from the history of malaria in a place and a time also means, for example, an understanding of DDT spraying as a human strategy that worked (1959) before it didn't (2009), and the role of climatic change and human political upheaval as enduring parts of our world. The dance will continue: human medicine and bureaucratic response will have to become coordinated partners in that moving complexity. The answer will be in continuing the dance.

Figure A.1. Rosa Haji Hamed. (Photo courtesy of Molly Williams.)

Rosa's Story

May 5, 2014

Rosa Haji Hamed survived her 1998 malaria ordeal that we visited in chapter 5. Her visit to the clinic had diagnosed her blood smear, and she received the medication Coartem from the clinic's store for falciparum malaria. She recovered. At five years of age she had won one of her life's early battles.

When I first met her in 2005 at her family home, she was a twelve-year-old young woman, approaching a marriageable age. Four years later, at the end of the Rockefeller maize-malaria project, she was an ambitious student, a young Muslim woman with expectations to continue her education. But life intervened; her father told me that he had a plan to betroth her to a wealthy local farmer. Islamic polygamy was not common in Waktola, but it was possible that she would become one of several wives in that rural though prosperous household.

In 2011, when I last met Rosa's loquacious father, I learned that she had won a victory of sorts. Her father, Haji Hamed, had relented and agreed to her marriage to a young town merchant (presumably handsome) who respected her educational ambitions. Rosa could continue her education in Asandabo's secondary school, located five kilometers from the house of her family and near the clinic she had visited as a five-year-old with symptoms of malaria.

Rosa's life and prospects were looking up. Asandabo had burgeoned as a thriving market town with a clinic and laboratory with rapid diagnostic technology (for malaria), a water tank (the water donkey could retire), and a link to the national power grid. One of Rosa's neighbors back in Waktola had a single electric socket that neighbors could use to charge their new mobile phones.

In July 2011 the Presidential Malaria Initiative (PMI) and USAID had marked their presence in Asandabo with a conspicuously large banner at the town center that announced their goal to control malaria as part of a national plan. In the local clinic, there was, however, no evidence of that plan.

We can assume that Rosa Haji Hamed still sees that banner a few times a week on her way to school.

Notes

Introduction: Malaria's Metaphor

Epigraph: Quoted in Russell E. Fontaine, Abdullah E. Najjar, and Julius S. Prince, "The 1958 Malaria Epidemic in Ethiopia," *American Journal of Tropical Medicine and Hygiene* 10 (1961): 795–803. "Bees in a smoked hive" is an apt metaphor quoted from a field report by Mogues, a young Ethiopian health worker describing his shock at the human suffering in a 1958 epidemic of falciparum malaria near the source of the Nile. In fact, smoke can kill bees in a hive, but when done carefully it can also serve to control them for honey collection.

1. Francis E. G. Cox, "History of the Discovery of the Malaria Parasites and Their Vectors," *Parasites and Vectors* 3 (2010): 3–5.

2. A movement in square dancing in which two dancers approach each other and circle back to back, then return to their original positions. The Middle French term is *dos à dos*, meaning "back to back."

3. Two compatible ideas about complexity and movement in ecology are those of Werner Heisenberg and C. S. Holling, "Understanding the Complexity of Economic, Ecological, and Social Systems," *Ecosystems* 4 (2001): 390–405.

4. Cox, "History of the Discovery," 1–2.

Chapter One: Ethiopia's Malaria in the Age of "the Bark"

1. Frederick Gamst, "A Note on a Malevolent Malaria Spirit and Its Significance to Public Health Workers," *Journal of Health* 6, no. 1 (1966): 24–25.

2. Asnakew Kebede, "Overview of the History of Malaria Epidemics in Ethiopia," paper presented at the Workshop on Capacity Building on Malaria Control, Addis Ababa, May 4–9, 2002; also see Gordon Covell, "Report on a Visit to Ethiopia September–October 1955," WHO7.0012 1946–1956. Covell was a malaria consultant to WHO's Eastern Mediterranean Regional Office.

3. Asnakew Kebede, "Overview of the History." See also K. Negash et al., "Malaria Epidemics in the Highlands of Ethiopia," *East African Medical Journal* 82, no. 4 (2005): 186–92.

4. Malaria fevers are notoriously misdiagnosed. But it is possible to do a retrospective historical diagnosis using seasons and geography of infection and to use older, but still accurate, spleen swelling exam counts to separate malaria outbreaks from typhus, relapsing fever, et cetera.

5. University of Hamburg, Digital Oriental Manuscript Library, Asien-Afrika-Institut, SM-011. See also British Library MS OR 534 f. 156r. These documents are in Ge'ez; their liturgical language offers no context for understanding *woba's* meaning. I am grateful to Magdelena Krzyzanowska of the Ludolf Centre, Hamburg University, and Brook Abdu of Boston University for these valuable references. The British Library manuscript describes a Persian invasion army's retreat after a malaria epidemic outbreak as it approached Ethiopia from the Saudi peninsula.

6. See James L. A. Webb Jr., *Humanity's Burden: A Global History of Malaria* (New York: Cambridge University Press, 2009), 95–98.

7. Sociologist Donald Levine, in his classic work *Wax and Gold*, also identifies personality as one of the primary attributes that highlanders claimed as a virtue: They viewed the lowlanders (*qolegnoch*) as heavy drinkers and hot-tempered; and highlanders (*degegnoch*) saw themselves as heavy eaters, and more inclined to litigation than to violence. Levine, *Wax and Gold: Tradition and Innovation in Ethiopian Culture* (Chicago: University of Chicago Press, 1972), 77–78.

8. Many areas of nineteenth-century Europe and North America suffered from malaria, though in its less deadly *vivax* form. Africa's predominant *falciparum* type was more deadly.

9. Fontaine, Najjar, and Prince, "The 1958 Malaria Epidemic in Ethiopia," 795–803. The conditions of those epidemics near the lake included early and prolonged rains and warmer temperatures. There are five types of human malaria, but only *P. falciparum* and *P. vivax* are relevant to Ethiopia and Africa as a whole.

10. James Bruce, *Travels to Discover the Source of the Nile, in the Years 1768, 1769, 1770, 1771, 1772, and 1773*, 5 vols. (Edinburgh: Ruthven, 1790). The American founding father John Adams (who owned a copy of Bruce's six-volume work), then at the Dutch court in Amsterdam, took "the bark" for his lingering fever in 1780. He was treated for the symptoms and recovered. David McCullough, *John Adams* (New York: Simon and Schuster, 2001), 264–65.

11. Malaria requires human bodies as well as mosquitoes to transmit itself. At the lakeside area of Ethiopia there was a small group of fisherfolk, the Wayto, who no doubt had acquired immunity from constant contact with malaria, but whose blood still held and transferred parasites to newcomers.

12. Henry Salt, *A Voyage to Abyssinia, and Travels into the Interior of That Country* (London: Rivington, 1814), 213; William C. Harris, *The Highlands of Ethiopia*, 3 vols. (London: Longman, 1844), 2:226, 3:268.

13. Ferret and Galinier are quoted in Richard Pankhurst, *Economic History of Ethiopia, 1800–1935* (Addis Ababa: Haile Sellassie I University Press, 1968), 629. For the sequence of malaria biomedical revelations, see Cox, "History of the Discovery," 3–4; and Randall M. Packard, *The Making of a Tropical Disease: A Short History of Malaria* (Baltimore: Johns Hopkins University Press, 2007), 111–36.

14. Henry A. Stern, *Wanderings among the Falashas in Abyssinia: Together with a Description of the Country and Its Various Inhabitants* (London: Wertheim, Macintosh, and Hunt, 1862), 151, 299. See also Richard Pankhurst, *Some Factors Influencing the Health of Traditional Ethiopia* (Addis Ababa: Haile Sellassie University Press, 1966).

15. Charles Johnston, *Travels in Southern Abyssinia, through the Country of Adal to the Kingdom of Shoa during the Years 1842–43* (London: Madden, 1844), 2:75, 101.

16. Walter C. Plowden, *Travels in Abyssinia and the Galla Country: With an Account of a Mission to Ras Ali in 1848* (London: Longmans, Green, 1868), 243–44.

17. Emilius A. De Cosson, *The Cradle of the Blue Nile: A Visit to the Court of King John of Ethiopia* (London: Murray, 1877), 2:299.

18. Samuel W. Baker, *The Nile Tributaries of Abyssinia and the Sword Hunters of the Hamran Arabs* (London: Macmillan, 1874), 46–48.

19. Kala-azar (or leishmaniasis) is also vector-borne and endemic to lowland zones, but its symptoms are skin ulcerations and not recurrent fever.

20. Mansfield Parkyns, *Life in Abyssinia: Being Notes Collected During Three Years' Residence and Travels in That Country* (London: Murray, 1853), 2:227. Louse and tick bites responsible for transmission of relapsing fever are less a seasonal phenomenon than the mosquito-borne falciparum malaria.

21. On louse-borne relapsing fever, see Sally J. Cutler, "Possibilities for Relapsing Fever Reemergence," *Emerging Infectious Diseases* 12, no. 3 (March 2006): 369–74. Nausea and vomiting ("bilious fever") are symptomatic of both falciparum malaria and relapsing fever—without blood smear slides, clinicians and local health workers often confuse the two—but the local recognition of *nidad* would have included both malaria types and relapsing fever.

22. Parkyns, *Life in Abyssinia*, 227.

23. Ibid., 286–87.

24. Stern, *Wanderings among the Falashas*, 229; Douglas C. Graham, *Glimpses of Abyssinia; or, Extracts from Letters Written while on a Mission from the Government of India to the King of Abyssinia in the Years 1841, 1842, and 1843* (London: Longmans, Green, 1867), 51.

25. Parkyns, *Life in Abyssinia*, 10–12.

26. Cox, "History of the Discovery," 2–6.

27. Richard Burton, *First Footsteps in East Africa; or, An Exploration of Harar*, ed. Gordon Waterfield (New York: Praeger, 1966), 53, 72.

28. Baker, *Nile Tributaries*, 148–49.

29. My thanks to Ato Sewnet Melaku, of the Bahir Dar Ministry of Health Malaria Unit, for his insight from his years as a Malaria Eradication Service technician. See chapter 6, "She Sings."

30. Asnakew Kebede, James C. McCann, Anthony E. Kiszewski, and Yemane Ye-Ebiyo, "New Evidence of the Effects of Agro-Ecological Change on Malaria Transmission," *American Journal of Tropical Medicine and Hygiene* 73, no. 4 (2005): 676–80.

31. De Cosson, *Cradle of the Blue Nile*, 2:213–14, 301.

32. Parisis, *L'Abisinnia* (Milan, 1888), 126, 128, 130, 133, 135–36.

33. Stern, *Wanderings among the Falashas*, 31. A lazaretto or "pesthouse" was a shelter of persons with infectious disease.

34. Baker, *Nile Tributaries*, 48–49.

35. Dermot R. W. Bourke, *Sports in Abyssinia; or, The Mareb and Tackazzee* (London: Murray, 1876), 130–31. "Chlordane" was a later term for an insecticide introduced in 1945. Bourke may have meant to say laudanum, a nineteenth-century cure-all for travelers.

36. Parkyns, *Life in Abyssinia*, 10–12.

37. Ibid.

38. Johnston, *Travels in Southern Abyssinia*, 126–27.

39. Ibid., 251–52, 262. Darwin's hydrotherapy was to relieve symptoms including agoraphobia and dizziness. William B. Bean, "The Illness of Charles Darwin," *American Journal of Medicine* 65, no. 4 (1978): 572–74. For Johnston, cupping involved the use of heated vials or horn cups to create a vacuum that drew blood to the fleshy places along the sufferer's body.

40. Johnston, *Travels in Southern Abyssinia*, 2:309; Plowden, *Travels in Abyssinia*, 108.

41. Johnston, *Travels in Southern Abyssinia*, 2:329–30.

42. Special thanks here to Prof. Jacques Mercier for his insights into *zar* possession and the role of the spirit world in disease amelioration.

43. This and the remaining excerpts in this chapter are from Jacques Mercier, *Asrès, le magicien éthiopien: Souvenirs 1895–1985* (Paris: Lattès, 1988), 384–91. My thanks to Michael Holm for his help in my translation of the French transcripts.

44. Here Asres's meaning is unclear. He might be referring to the use of incense or smoke from herbs or native wood, such as that used for smoking a beehive.

Chapter Two: Mindscapes of Malaria

Epigraphs: Charles Johnston, *Travels in Southern Abyssinia, through the Country of Adal to the Kingdom of Shoa Dduring the Years 1842–43* (London: Madden, 1844); Mario Giaquinto Mira, "La lotta antimalarica in A.O.I.," in *Opere per l'organizzazione civile in A.O.I.* (Addis Ababa: Servizio Tipografico Governo Generale A.O.I., 1939), 26–27.

1. These worlds overlapped: for example the origin and Charles Darwin's survival of the fittest principle came not to Darwin, but in 1858 to naturalist Alfred Russel Wallace during (as he later recalled) his fit of malarial fever on an island called Ternate in the Indonesian straits. See Simon Winchester, *Krakatoa: The Day the World Exploded, August 27, 1883* (New York: HarperCollins, 2003), 60.

2. Colleague Rich Pollack, malaria entomologist, tells me that he would not rule out the twenty-first century from this role of persistent beliefs. On the sanitarian mindset, see Nancy Tomes, *The Gospel of Germs: Men, Women, and the Microbe in American Life* (Cambridge, MA: Harvard University Press, 1998).

3. Randall M. Packard, *The Making of a Tropical Disease: A Short History of Malaria* (Baltimore: Johns Hopkins University Press, 2007), 19–35; James L. A. Webb Jr., *Humanity's Burden: A Global History of Malaria* (New York: Cambridge University Press, 2009), 42–65.

4. Packard, *Making of a Tropical Disease*, 31–53.

5. C. E. Bosworth, E. van Donzel, B. Lewis, and Ch. Pellat, eds., *The Encyclopaedia of Islam*, vol. 6 (Leiden: Brill, 1991), 229–30. See also A. R. Zahar, "Review of the Ecology of Malaria Vectors in the WHO Eastern Mediterranean Region," *Bulletin of the World Health Organization* 50, no. 5 (1974): 427–40; and J. de Zulueta, "Malaria and Mediterranean History," *Parassitologia* 15 (1973): 1–15.

6. British Library Ms. OR 534 f. 156r. See also the Ge'ez-language parchment manuscript shown in chapter 1. My thanks to Brook Abdu for pointing out this document.

7. Guidi's text uses Italian and Ge'ez script. Ignazio Guidi, *Vocabolario Amarico-Italiano* (Rome: Istituto per l'Oriente, 1935), 395, 578. My translation.

8. Tesema Hapta Mikael, author and translator *Yamarigna Mazgeb Qalat* (Addis Ababa, 1951 E.C., 1958–1959). My translation.

9. Thomas L. Kane, *Amharic-English Dictionary* (Wiesbaden: Harrassowitz, 1990), 1:1056–57.

10. Ibid., 2:1535.

11. Wolf Leslau, *Concise Amharic Dictionary* (Berkeley: University of California Press, 1996), 171.

12. My thanks to Getahun Mesfin and Grover Hudson at Michigan State University for their help in compiling these dictionary references from the excellent Michigan State library collection.

13. Aguë, "acute (fever)" (Modern French *fièvre aigüe*) refers to symptoms of recurrent fever—probably malaria—in European folk diagnoses.

14. *Matshaf Senksar baGe'ezna BaAmarigna kaMegabit iska Pagume*. This volume is undated, but held in the collection of the Ludolf Centre at the University of Hamburg.

15. Andrew J. Carlson and Dennis G. Carlson, *Health, Wealth, and Family in Rural Ethiopia: Kossoye, North Gondar Region, 1963–2007* (Addis Ababa: Addis Ababa University Press, 2008), 93–94. In 1965, WHO recorded, optimistically, 81 hospitals, 57 health centers, and 432 health stations in Ethiopia. See Plan of Operation for a Pre-Eradication Program in Ethiopia, January 18, 1965, Appendix IV, Annex I. WHO M2/372/3 (A) ETH.

16. Frank M. Snowden, *The Conquest of Malaria: Italy, 1900–1962* (New Haven, CT: Yale University Press, 2006), 38–45; Giaquinto Mira, "La lotta antimalarica."

17. Snowden uses the term "magic bullet." Snowden, *Conquest of Malaria*, 44–51; Packard, *Making of a Tropical Disease*, 121, 128–31. Insecticide spray containing DDT would be the silver bullet after World War II.

18. Snowden, *Conquest of Malaria*, 95–101.

19. Ibid., 128, 138–39.

20. After false claims of effectiveness, the malaria specialist Alberto Missiroli repeated the test on Sardinian peasants and found no value to the mercury. Snowden, *Conquest of Malaria*, 145–46.

21. Snowden refers to this as "biological warfare." Snowden, *Conquest of Malaria*, 187–90. He argues that it was a specific and deliberate plan. Elizabetta Novello, historian of malaria at the University of Padua, disagrees and has told me that malaria's spread after the German retreat was an "unintended consequence" of the destruction of the local irrigation.

22. Paris Green was, in fact, the best substance available at the time. It may have saved many lives, for a time. It was copper acetoarsenite, a highly toxic emerald-green powder used as a larvicide, insecticide, fungicide, pigment, and wood preservative. It was used extensively for banana cultivation in Central America. Workers' skin turned blue-green and they were called "parakeets." Paris Green's cost and toxicity, however, ultimately made it unsuitable for large-scale antimalaria spraying campaigns.

23. Augusto Corradetti, "Le conoscenze sulla distribuzione delle species Anofeliche nell'Africa Orientale Italiana," *Rivista di Biologia Coloniale* 3, no. 4 (1940): 419–29, found in WHO7.001 JKT 1 Ethiopia 1937–1942.

24. Giovanni B. Grassi, *Studi di uno zoologo sulla malaria* (Rome: Accademia dei Lincei, 1900); Francis E. G. Cox, "History of the Discovery of the Malaria Parasites and Their Vectors," *Parasites and Vectors* 3 (2010): 3–6; and Webb, *Humanity's Burden*, 135–36. Also see Snowden, *Conquest of Malaria*, 181–97.

25. Snowden, *Conquest of Malaria*, 30, 220; and Haile M. Larebo, *The Building of an Empire: Italian Land Policy and Practice in Ethiopia, 1935–1941* (New York: Oxford University Press, 1994), 65–70.

26. In fact, historian Elisabetta Novello has pointed out that farmers and workers from these regions may well have had genetic traits for sickle-cell anemia (Anemia mediterranea) and thus resistance to malaria, giving them some protection that they may have carried to Ethiopia. Personal communication, 2013.

27. Few reports identified malaria as part of this failure, but the agrogeography of failure in the Italian period duplicated the challenges that faced Ethiopia's own postwar planning in the 1950s where malaria played a role. See Larebo, *Building of an Empire*, 291–95.

28. See the 1940 English-language Italian-propaganda publication *Main Features of Italy's Action in Ethiopia, 1936–1941* (Florence: Istituto Agricolo Coloniale, 1946); and James C. McCann, *People of the Plow: An Agricultural History of Ethiopia, 1800–1990* (Madison: University of Wisconsin Press, 1995).

29. Those transplanted peasant farmers showed typical pragmatism. After a year's forced experimentation in sowing Italian improved wheat seeds, most of the Italian farmers planted their own *granoturco* (maize), a pattern that had taken place three centuries earlier in Veneto, Rovigo, and Treviso when the Venetian state had tried to force the cultivation of wheat. See James C. McCann, *Maize and Grace: Africa's Encounter with a New World Crop, 1500–2000* (Cambridge: Harvard University Press, 2009).

30. Giaquinto Mira, "La lotta antimalarica."

31. Ibid., 26–27.

32. This map from Italian malaria specialists includes sites primarily in Eritrea and the Awash Valley. The Lake Tana Basin and areas in the southwest were surveyed later.

33. Presumably the program was for Italian settlers and agents, though he does not say so. Castellani summary of Corredetti Report to Office International d'Hygiene Publique, October 18, 1938. WHO Archive, Ethiopia 1937–1942, JKT1 WHO 7.0011.

34. Gordon Covell, "Malaria in Ethiopia," *Journal of Tropical Medicine and Hygiene* 60 (January 1957): 7–16. This report contains evidence gathered in his earlier 1952 field visit to the Lake Tana region.

35. Gordon Covell's field report from 1957 and an earlier spotty and general survey by British specialists from 1945 reported on mosquito types in certain locations. A. R. Melville et al., "Malaria in Abyssinia," *East African Medical Journal* 22 (September 1945): 285–94.

36. Ibid., 293.

Chapter Three: Flight of the Valkyries

Epigraphs: James Bruce, *Travels to Discover the Source of the Nile, in the Years 1768, 1769, 1770, 1771, 1772, and 1773* (Edinburgh: Ruthven, 1790), 4:8–9: and Jacques Mercier, *Asrès, le magicien éthiopien: Souvenirs 1895–1985* (Paris: Lattès, 1988), 388–89.

1. Tana, however, is not technically a pulse lake, like Cambodia's Tonle Sap on the Mekong River or Botswana's vast Okivongo wetland. The Blue Nile's waters rise and fall with the seasons, whereas the Tonle Sap, for example, has an annual backflow that sustains its distinctive aquatic ecology.

2. Henry Lamb et al., "Late Pleistocene Desiccation of Lake Tana, Source of the Blue Nile," *Quaternary Science Reviews* 26, nos. 3–4 (2007): 287, 296.

3. Abebe Getahun and Eshete Dejen, *Fishes of Lake Tana: A Guide Book* (Addis Ababa: Addis Ababa University Press, 2012). So far humans have not tried to add new fish species to Tana, as they have with the Nile perch in Lake Victoria.

4. Another mosquito species found in the puddles and pools, *An. arabiensis*, has another story (see below).

5. William C. Harris, *The Highlands of Ethiopia* (London: Longman, 1844), 3:265.

6. Bruce, *Travels to Discover the Source of the Nile*. For a lively account of Bruce's audience reception, see Alan Moorehead, *The Blue Nile* (London: Hamish Hamilton, 1962), 22–51.

7. For these episodes colorfully described, see Moorehead, *Blue Nile*, 26–35. But Moorehead wrongly notes the White Nile as the source of the Nile waters that reach Cairo. Over 80 percent of the waters reaching Cairo flow from the Ethiopian highlands, and the Blue Nile Basin is the primary source.

8. R. K. Pachauri and A. Reisinger, eds., *Climate Change: Synthesis Report* (Geneva: IPCC, 2007).

9. Spleen rates are a rough measure of endemic malaria in a certain site. Researchers examine and register the percentage of enlarged spleens in children two to six years of age. In areas of persistent malaria, the rates would be 75 percent and above. In areas with less exposure or unstable transmission, the rates would be lower. High spleen rates would indicate acquired immunity, which was rare in Ethiopia's highlands.

10. A. R. Melville et al., "Malaria in Abyssinia," *East African Medical Journal* 22 (September 1945): 287, 293.

11. A Mobile Malaria Section of the British East African Medical Corps carried out a malaria survey between April 1941 and March 1942. The wide-ranging report commented on the general ecology of malaria in Abyssinia (Ethiopia), but not on the Blue Nile watershed. See chapter 4 for details on its work in the Jimma area. Melville et al., "Malaria in Abyssinia," 285–93.

12. Andrew J. Carlson and Dennis G. Carlson, *Health, Wealth, and Family in Rural Ethiopia: Kossoye, North Gondar Region, 1963–2007* (Addis Ababa: Addis Ababa University Press, 2008), 15–17.

13. Sir Gordon Covell, "Malaria in Ethiopia," *Journal of Tropical Medicine and Hygiene* 60 (January 1957): 7.

14. Ibid., 11–12.

15. Ibid., 11.

16. Asnakew Kebede, personal communication, May 2013. Asnakew is a native son of Kolladuba, on the way from Gondar to the lake.

17. Mercier, *Asrès*, 384. Mercier's account is a verbatim translation of Asres's recorded oral narration, his lively memory.

18. Russell E. Fontaine, Abdallah E. Najjar, and Julius S. Prince, "The 1958 Malaria Epidemic in Ethiopia," *American Journal of Tropical Medicine and Hygiene* 10, no. 6 (1961): 797. See also "Report on the Second Regional Conference on Malaria Eradication," Addis Ababa, November 16–21, 1959, in WHO. Also WHO file EM/ME Tech. 2/1-54, 1959.

19. Fontaine, Najjar, and Prince, "1958 Malaria Epidemic in Ethiopia," 800.

20. Death rates may have been especially high given that the 1957 harvest year in the northeast of the country had seen a severe famine. See James McCann, *From*

Poverty to Famine in Northeast Ethiopia: A Rural History, 1900–1935 (Philadelphia: University of Pennsylvania Press, 1987), 199.

21. Fontaine, Najjar, and Prince, "1958 Malaria Epidemic in Ethiopia," 795–803.

22. Ibid., 802.

23. Ibid.

24. Ibid., 797.

25. George Davidson, "*Anopheles gambiae*: A Complex of Species," *Bulletin of the World Health Organization* 31, no. 5 (1964): 625–34. He also noted unassumingly: "From a practical point of view it may be necessary for the field worker to be able to identify the exact species with which he is dealing before the most efficient means of controlling it can be found."

26. I have often asked my entomology colleagues, who have decades of research experience on malarial mosquitoes, how the insects survive dry conditions and then recover. They respond that it continues to be a mystery. Dr. Rebecca Robich, personal communication.

27. Davidson, "*Anopheles gambiae*," 634.

28. "Quarterly Report Covering Period 1 October to 32 December 1965," WHO archive M2/372/3 (b) ETH. Another batch of eggs had been sent to the Sperimentali di Entomologia in Manticelli, Italy.

29. Yilma Mekuria and Girma WoldeTsadik, "Malaria Survey in North and North Eastern Ethiopia," *Ethiopian Medical Journal* 8 (1970): 201–6.

30. Annual Report Malaria Eradication Program in Ethiopia July 1968–July 1969, WHO Archive, M2/374/3 (b) ETH, 6–7.

31. "Information on Entomological Activities of the National Malaria Eradication Service of the Imperial Ethiopian Government," WHO7.0013 ETH 1972, Annex No. 2 (Species Distribution).

Chapter Four: Tragedy of the Jeep, 1958–1991

1. WHO Quarterly Report, 1971, WHO archive M2/372/3; WHO7.003 ETH 1970–1971, p. 2. The plan was signed in 1957. An addendum was signed in 1958 to record the agreement on the establishment of the malaria training center at Nazaret (now Hadama).

2. Russell E. Fontaine, Abdallah E. Najjar, and Julius S. Prince, "The 1958 Malaria Epidemic in Ethiopia," *American Journal of Tropical Medicine and Hygiene* 10, no. 6 (1961): 795–803.

3. WHO7 0013 ETH 1970–1971, 1–2.

4. Wikipedia, s.v. "DDT," http://en.wikipedia.org/wiki/DDT.

5. A. Najjar and Russell Fontaine, Dembia Pilot Project, Beghemder Province, Ethiopia, U.S.I.C.A., October 22, 1959, 9. This report indicates that thirty-one susceptibility tests were carried out in Ethiopia prior to 1959 on the three major vector mosquito species. For the original report, see Pierre H. A. Joliet, *Observations of the Ethiopian Anopheles Mosquitoes and Their Susceptibility to Insecticides*, Second Regional Conference on Malaria Eradication, Addis Ababa, November 16–21, 1959, EM/NE–Tech.2/23, October 26, 1958.

6. In 1971, the parasite formula used by WHO stated that in Ethiopia malaria cases showed 60 percent to 75 percent *P. falciparum*, 19 percent to 25 percent *P. vivax*, and 6 percent to 15 percent *P. malariae*. Mixed vivax and falciparum cases were common.

WHO 7.0013 ETH 1970–1971. *P. vivax*, the less deadly but recurrent fever-producing malaria types, seemed to be less present during severe epidemics like 1953.

7. Report on the Second Regional Conference on Malaria Eradication, Addis Ababa, November 16–21, 1959. WHO Reference Library, Geneva (EM/MAL/38–48, 1959–1963).

8. By 1991, the operative figure for types of anopheline mosquitoes was forty-two. All vector inventories up to 1991 agreed that *An. gambiae s.l. (An. arabiensis)* was the main vector, with *An. funestus* and *An. pharoensis* and *A. nili* as secondary vectors. Names for mosquito species now use *An.* rather than *A.* to name species. Richard Pollack, personal communication, 2013.

9. Frank M. Snowden, *The Conquest of Malaria: Italy, 1900–1962* (New Haven, CT: Yale University Press, 2006), 109, 188–89.

10. Joliet, *Observations*, 13.

11. WHO7.0013 ETH 1970–1971, 6.

12. http://en.wikipedia.org/wiki/Sulfonamide_%28medicine%29#Side_effects.

13. Aldo Castellani, Delegue des Colonies Italiennes, Epidemiologie du Paludism dans la Region du Lac Tsana, WHO7 0011 Archive, Ethiopia 1937–1942.

14. See James A. Michener, *Tales of the South Pacific* (New York: Macmillan, 1947), 165–70. My father served as a medical orderly at Guadalcanal, New Caledonia, and New Britain in 1942–1943.

15. Other drugs became available in world markets, including a pyrimethamine sulfadoxine combination, mefloquine, halofantrine, primaquine, and quinine injections when chloroquine failed. Awash Tekle Haimanot, Report on Visit to Ministry of Health, Addis Ababa, Ethiopia and Field Visits. WHO7 0014. ETH 1991.

16. Awash Tekle Haimanot, Report on Visit to Ministry of Health, Addis Ababa, Ethiopia and Field Visits. WHO7 0014. ETH 1991, p. 2.

17. Andrew J. Carlson and Dennis G. Carlson, *Health, Wealth, and Family in Rural Ethiopia: Kossoye, North Gondar Region, 1963–2007* (Addis Ababa: Addis Ababa University Press, 2008).

18. The official terminology at WHO and ICA had been to call the stage of mapping and spray pilot programs pre-eradication, an early effort that sought to bring African sites, like Ethiopia, up to speed with other world areas.

19. Official Record of the World Health Organization, Geneva (MAL vol. 176), 18–19, 120 (my italics).

20. Ibid.

21. Ibid., 18–19.

22. Dr. S. C. Edwards, Report on a Visit to Ethiopia, December 10–18, 1969, WHO7.0012, ETH 1968–1969, January 1970.

23. Report by a Strategy Review Team, Ethiopia, May 6–27, 1970. WHO7.0013. WHO 1970–1971.

24. H. M. S. Morin, Malaria Problems in Eastern Mediterranean Region Preliminary Remarks, March 1951. World Health Organization Regional Office for the Eastern Mediterranean (EM/MAL/EM/MAK17) 1951–1955, 4. WHO library, Geneva.

25. The author visited this hospital in 1992 to deliver the injured from a road accident (an overturned flatbed truck of workers), victims with bleeding head trauma and broken limbs. The once proud hospital had no staff, medicines, or patient beds. A pedestrian we asked for directions to the hospital waved his hand in the direction he had come from and said "back there." We found out later he was the hospital's only doctor.

26. Awash Tekle Haimanot, Report on Visit to Ministry of Health, Addis Ababa, Ethiopia and Field Visits. WHO7 0014. ETH 1991. Awash Tekle Haimanot, Report on Visit to Ministry of Health, Addis Ababa, Ethiopia and Field Visits. WHO7 0014. ETH 1991, 2.

Chapter Five: Malaria Modern

Epigraph: James C. McCann, *Maize and Grace: Africa's Encounter with a New World Crop, 1500–2000* (Cambridge: Harvard University Press, 2009).

1. Gordon Covell, "Malaria in Ethiopia," *Journal of Tropical Medicine and Tropical Hygiene* 60 (January 1957): 13–14. In the 1950s, WHO suggested using the spleen rates (percent of children with enlarged spleens) as a proxy of malaria endemicity. A rate of 50 percent enlarged spleens meant "hyperendemicity," in other words, seasonal outbreaks but no acquired immunity, as in most of Ethiopia; 75 percent meant "hyperendemicity" and immunity among those children who survived, a rarity in Ethiopia. C. P. Coogle, "The Spleen Rate as a Measure of Malaria Prevalence in the United States," *Public Health Reports* 42, no. 25 (1927): 1683–88.

2. The most recent research on the effect of eucalyptus on water tables is Tilashwork Chanie Alemie, "The Effect of *Eucalyptus* on Crop Productivity and Soil Properties in Koga Watershed, Western Amhara Region, Ethiopia" (PhD Thesis, Cornell University, 2009). Her evidence shows an effect on soil dryness, chemistry, and maize development within 15 meters of eucalyptus rows, but none at 40 meters. Koga is 30 kilometers from Bahir Dar town.

3. Bekele Abebe operated a private pharmacy and was Burie district's only healthcare provider beginning in 1968 until 2000. He reports that malaria cases in the district were extremely rare and then consisted only of people who sought his care from malaria endemic lowlands. Interview with Bekele Abebe, Addis Ababa, May 24, 2003. When I lived in Burie from 1973 to 1975, there were no mosquitoes at dusk or at night.

4. Interview with Semahagne Abate, head of the Burie Anti-Malaria Association, June 2000.

5. My partner in this research has been Asnakew Kebede of the Ethiopian Ministry of Health, who holds an MSc in epidemiology from the London School of Hygiene and Tropical Medicine. His assiduous compiling of local case data, curiosity, and participation in daily brainstorming sessions has been essential to this research.

6. Three published studies of laboratory work on mosquito links to maize pollen now exist. See Yemane Ye-Ebiyo, Richard J. Pollack, and Andrew Spielman, "Enhanced Development in Nature of Larval *Anopheles Arabiensis* Mosquitoes Feeding on Maize Pollen," *American Journal of Tropical Medicine and Hygiene* 63, nos. 1–2 (2000): 90–93; Yemane Ye-Ebiyo, Richard J. Pollack, Anthony Kiszewski, and Andrew Spielman, "Enhancement of Development of Larval *Anopheles Arabiensis* by Proximity to Flowering Maize (*Zea Mays*) in Turbid Water and When Crowded," *American Journal of Tropical Medicine and Hygiene* 68, no. 6 (2003): 748–52; Yemane Ye-Ebiyo, Richard J. Pollack, Anthony Kiszewski, and Andrew Spielman, "A Component of Maize Pollen That Stimulates Larval Mosquitoes (Diptera: Culicidae) to Feed and Increases Toxicity of Microbial Larvicides," *Journal of Medical Entomology* 40, no. 6 (November 2003): 860–64.

7. Ameneshewa B. and M. W. Service, "The Relationship between Female Body Size and Survival Rate of the Malaria Vector *Anopheles arabiensis* in Ethiopia," *Medical Veterinary Entomology* 10, no. 2 (1996): 170–72.

8. Professor Spielman approached me, having heard of my research project on maize in Africa. I then proposed the research site at Burie and he arranged for me to meet my local research partner, Asnakew Kebede.

9. In May 2003, I returned to Burie district to gather further case evidence of malaria and to assess the impact of a rare failure of the main rains in the previous season. This drought was the first in the West Gojjam region for many years. Collecting and assessing climate data presented a real problem since the National Meteorological Service had closed its Burie data collection station some years before. We therefore took an alternative strategy by asking farm unit managers at the adjacent privatized Bir Sheleko state farm. They graciously provided data from their farm site that runs parallel to the border with Burie, five kilometers to the east. This data provided us with proxy information about Burie's daily temperature and rainfall for the period 1997–2000. The Bir Sheleko farm is in a flat-bottomed valley ideally suited for its large-scale mechanized maize production operation. Its raw figures for temperature and rainfall would differ somewhat from the adjacent Burie highlands, but the annual patterns, we argued, would be consistent.

10. I am grateful to Dr. Abdusamad Hajj Ahmed, a native son of Burie, for sharing his trading family's experience with malaria with me.

11. Interview with Getaneh Atras, age fifty-three, Imbodbod (Gulim), May 24, 2002.

12. See James C. McCann, *People of the Plow: An Agricultural History of Ethiopia, 1800–1990* (Madison: University of Wisconsin Press, 1995), chap. 8. The late Ian Watt, geographer at Addis Ababa University, was the first person to call my attention to maize's growing role in Ethiopian agriculture.

13. See James Keeley and Ian Scoones, "Knowledge, Power and Politics: The Environmental Policy-Making Process in Ethiopia," *Journal of Modern African Studies* 38, no. 1 (2000): 97–101.

14. Market observations in May 2002 indicated that older, multicolored composite varieties were still grown, but in much smaller amounts than BH660. Large grain merchants accepted only the pure white BH660 maize, while the older types were sold in smaller amounts for local consumption and for seed.

15. Asnakew Kebede, James C. McCann, Anthony E. Kiszewski, and Yemane Ye-Ebiyo, "New Evidence of the Effects of Agro-Ecologic Change on Malaria Transmission," *American Journal of Tropical Medicine and Hygiene* 73, no. 4 (2005): 676.

16. Waktola had similar numbers, though maize was more fully grown as a field crop there.

17. This figure is based on informal field surveys conducted in May 2002 in three sites. In each of these places, maize cultivation was especially concentrated around new housing sites.

18. Interview with Getaneh Atres, Imborbor (Gulim), May 24, 2002.

19. For "high modernism," see epilogue; and James C. Scott, *Seeing Like a State: How Certain Schemes to Improve the Human Condition Have Failed* (New Haven, CT: Yale University Press, 1998).

20. Interview with Getaneh Atres (Gulim), May 24, 2002.

21. Gojjam's nongovernmental Anti-Malaria Association unofficially tallied 222,992 cases and 7,783 deaths, but estimated the actual numbers were twice that amount

(personal communication from Semahagne Abate, Burie secretary of the Anti-Malaria Association).

22. The larger size and greater longevity for maize pollen–fed *Anopheles arabiensis* suggests a great extrinsic life-cycle completion rate for the parasite and therefore a higher rate of transmission (i.e., more, longer-lived mosquitoes carrying more deadly parasites).

23. The Ministry of Agriculture taxonomy agreed with our own assessment of the areas we assessed directly and their involvement in maize production.

24. That report, coauthored by James McCann and Richard Pollack of Boston University and the Harvard School of Public Health, now appears in published form as a working paper in the Program for the Study of African Environmental History. The effort described there results from the collaborative and productive efforts of scholars at Boston University, Harvard University, Bentley University, the World Health Organization, the Ethiopian Ministry of Health's Malaria Unit, the Ethiopian Ministry of Agriculture's Ethiopian Institute of Agricultural Research, Addis Ababa University, Jimma University, and Wollega University, as well as with colleagues from the Massachusetts Institute of Technology and the University of Vermont.

25. The polymerase chain reaction (PCR) is a biochemical technology in molecular biology used to amplify a single or a few copies of a piece of DNA across several orders of magnitude. This analysis was done in the laboratory at the Harvard School of Public Health to identify species of mosquito.

26. Owners of the fields of detasseled maize were compensated for any crop losses that might have resulted from reduced pollination and for the use of their fields as study sites. In collaboration with colleagues from the Ethiopian Institute for Agricultural Research who know the farm community well, we developed a method for pollen control/management. We based our estimates of the prices of the crop during the previous growing season and the extent of the planting during the current season.

27. This measure relies on the MacDonald formula, in which the most important element of malaria transmission is mosquito population. See George MacDonald, *The Epidemiology and Control of Malaria* (Oxford: Oxford University Press, 1957).

28. Ameneshewa and Service, "Relationships between Female Body Size and Survival Rate," 170–72.

29. I owe special thanks to Dr. Twumasi Afriye and Dr. Benti Tolossa, both experienced maize breeders who met with me in Addis Ababa and at CIMMYT headquarters in El Batan, Mexico, in 2003 and 2005.

30. Ameneshewa and Service, "Relationships between Female Body Size and Survival Rate," 170.

31. Technical details: A total of 542 third instar larvae were assessed in this manner: 242 from breeding sites in close proximity to tasseled maize plants, 121 from detasseled sites, and 179 from sites where a barrier crop (usually peppers) replaced the maize normally cultivated in plots proximal to breeding sites. Larvae from tasseled sites survived the longest (5.01 days ±1.64), as compared to detasseled sites (4.88 days ±1.66) and barrier sites (4.43 days ±1.37). Paired T-tests revealed significant differences between those derived from barrier and tasseled sites ($T = -3.98$; $P<0.001$), and barrier and detasseled sites ($T = -2.48$; $P=0.014$), but not between tasseled and detasseled sites ($T = 0.71$; $P = 0.473$).

32. More relevant details: The soluble content of maize pollen (BH660) was extracted with deionized water and 80 percent ethanol, and analyzed for total protein

(Kjeldahl method), soluble protein (Lowry method), total fat (Soxhlet extraction method), total carbohydrate (phenol–sulfuric acid method), total ash content (blast furnace), and total moisture (oven heating). Surprisingly, no prior study of pollen had been done until this project that sought to understand the connection between larval nutrition and survival. See Bezawit Eshetu, "Nutrient Composition of Maize Pollen and Its Microbial Degradation" (MSc thesis, Biology Department, Addis Ababa University, 2007). Here are the formal details: Of the total carbohydrate and protein components, we measured 24 milligrams soluble protein, 179 milligrams soluble carbohydrate (extracted by 80 percent ethanol), 153 milligrams soluble carbohydrate (extracted by deionized water), and 51 milligrams of starch per 1 gram of pollen.

33. Populations of bacteria and yeasts were detected in greater densities in habitats where maize pollen was freely shed. The bacterial population in the habitat was dominated by species of *Bacillus, Pseudomonas, Micrococcus,* and *Serratia.* The common molds in the water samples were *Aspergillus, Fusarium, Stemphylium, Microascus, Verticillium, Gelasinospora,* and *Botrytis.*

34. Special thanks to coauthors Richard Pollack, Anthony Kiszewski, Zelalem Teffera, Melaku Wondafrash, and Michael Raviesi for their fieldwork on this topic.

35. Culicines (or *Culex*) were the most common and abundant of all mosquito larvae sampled, found in 62 percent of all borrow pits, and occurred together with *An. arabiensis* in 29 percent of habitats sampled.

36. Maize is the predominant plant in the area that utilizes a "C4" photosynthetic mechanism. Most other plants in the Ethiopian agricultural landscape—grains, tubers, beans, peas, peppers—are "C3" plants.

37. Magaly Koch and James C. McCann, "Satellite Imagery, Landscape History, and Disease: Mapping and Visualizing the Agroecology of Malaria in Ethiopia" (Working Paper 8, PSAE Research Series, Boston University, 2010).

38. Michael R. Reddy et al., "Outdoor Host Seeking Behaviour of Anopheles gambiae Mosquitoes Following Initiation of Malaria Vector Control on Bioko Island, Equatorial Guinea," *Malaria Journal* 10 (2011):184. In our Waktola project area, we surveyed bednets actually *used* nightly rather than by the numbers distributed. Only about a third of rural households in our Waktola survey actually used bednets for mosquito protection. Others used them as oxen tethers, building materials, or clothing.

39. The study is described in the following article: Ryan E. Trudel and Arne Bomblies, "Larvicidal Effects of Chinaberry (*Melia azederach*) Powder on *Anopheles arabiensis* in Ethiopia," *Parasites and Vectors* (2011): 4:72.

Chapter Six: She Sings

Epigraphs: Charles T. O'Connor Jr., "The Distribution of Anopheline Mosquitoes in Ethiopia," *Mosquito News* 27, no. 1 (March 1967): 44; G. B. White, "Anopheles Gambiae Complex and Disease Transmission in Africa," *Transactions of the Royal Society of Tropical Medicine and Hygiene* 68, no. 4 (1974): 293–94.

1. Thierry Lefèvre et al., "Beyond Nature and Nurture: Phenotypic Plasticity in Blood-Feeding Behavior of *Anopheles gambiae* s.s. When Humans Are Not Readily Accessible," *American Journal of Tropical Medicine and Hygiene* 81, no. 6 (2009): 1023–29. Here the use of the term "plasticity" implies flexibility, movement, and adaptation to circumstances—always amenable to change.

2. See http://answers.yahoo.com/question/index?qid=20090809122823AAJ VBO4.

3. Bert Hölldobler and Edward O. Wilson, *The Superorganism: The Beauty, Elegance, and Strangeness of Insect Societies* (New York: Norton, 2009), 5–10.

4. J. D. Charlwood et al., "The Rise and Fall of *Anopheles arabiensis* (Diptera: Culicidae) in a Tanzanian Village," *Bulletin of Entomological Research* 85, no. 1 (1995): 42. *An. gambiae* Species A are the best malaria carriers, since she bites only humans (i.e., anthropohilic biting).

5. Ronald Ross, *The Prevention of Malaria (with Addendum on the Theory of Happenings)*, 2nd ed. (London: Murray, 1911).

6. Bert Hölldobler and Edward O. Wilson, *The Ants* (Cambridge, MA: Harvard University Press, 1990).

7. Edward O. Wilson, "Karl Marx Was Right, Socialism Works," unpublished interview by Frans Roes, Harvard University, March 27, 1997. The full text can be found at http://www.froes.dds.nl/WILSON.htm.

8. Bert Hölldobler and Edward. O. Wilson, *The Leafcutter Ants: Civilization by Instinct* (New York: Norton, 2011), 89–91.

9. Ibid., 91.

10. See, for example, Augusto Corradetti, "Notizie preliminari sulla fauna Anofelica della regione di Gondar del Lago Tana e della regione del Simien," *Rivista di Parassitologia* 3, no. 2 (1939): 153–56; and Mario Giaquinto Mira, " La lotta antimalarica in A.O. I.," in *Opere per l'organizzazione civile in Africa orientale italiana* (Addis Ababa: Servizio Tipografico Governo Generale A.O.I.,1939). Note that the A. here means *anopheline*, where An., used later, was a way of distinguishing *anopheline* from *aedes*, including the vector A. (*aedes*) *aegypti*, the genus and species of tropical mosquitoes that transmit yellow fever and dengue.

11. George Davidson, "*Anopheles gambiae* Complex," *Nature* 196 (1962): 907.

12. George Davidson, "*Anopheles gambiae*: A Complex of Species," *Bulletin of the World Health Organization* 31, no. 5 (1964): 625–34.

13. Ibid.

14. O'Connor, "Distribution of Anopheline Mosquitoes in Ethiopia," 42–54.

15. A. Parmakelis et al., "Historical Analysis of a Near Disaster: *Anopheles gambiae* in Brazil," *American Journal of Tropical Medicine and Hygiene* 78, no. 1 (2008): 176–78. Actually, Rich Pollack has pointed out that malaria in that Brazilian valley had been minor since the local mosquitoes were poor vectors. When *An. arabiensis* arrived on a French ship, it changed the local malaria equation and the sudden outbreak occurred. Rich said it had an effect like "a baseball player on steroids."

16. Delenasaw Yewhalaw et al., "First Evidence of High Knockdown Resistance Frequency in *Anopheles arabiensis* (Diptera: Culicidae) from Ethiopia," *American Journal of Tropical Medicine and Hygiene* 83, no. 1 (2010): 122–25; Meshesha Balkew et al., "Insecticide Resistance in *Anopheles arabiensis* (Diptera: Culicidae) from Villages in Central, Northern, and South West Ethiopia and Detection of *kdr* Mutation," *Parasites and Vectors* (2010): 3:40.

17. For the contrast with ant social behavior, see Hölldobler and Wilson, *Ants*.

18. Rope-a-dope: *n.* (Individual Sports & Recreations / Boxing): (a) a method of tiring out a boxing opponent by pretending to be trapped on the ropes while the opponent expends energy on punches that are blocked; (b) (*as modifier*) rope-a-dope strategy. (Coined by U.S. boxer Muhammad Ali.) *Collins English*

Dictionary—Complete and Unabridged, s.v. "rope-a-dope," http://www.thefreedictionary.com/rope-a-dope.

Epilogue: The Dance Continues

1. On high modernism and complexity, see James C. Scott, *Seeing Like a State: How Certain Schemes to Improve the Human Condition Have Failed* (New Haven, CT: Yale University Press, 1998); and C. S. Holling, "Understanding the Complexity of Economic, Ecological, and Social Systems," *Ecosystems* 4 (2001): 390–405.

2. Leslie Roberts and Martin Enserink, "Did They Really Say . . . Eradication?" *Science* 318 (2007): 1544–45. See also the program for the February 2014 conference "The Science of Malaria Eradication," sponsored by the Gates Foundation. Papers presented showed a rather mixed message of expectations.

3. malEra Consultative Group on Monitoring, Evaluation, and Surveillance, "A Research Agenda for Malaria Eradication: Monitoring, Evaluation, and Surveillance," *PLoS Med* 8, no. 1 (2011), www.plosmedicine.org. Here is the argument for surveillance made by the Carter Center, though it inexplicably continues to use the catchword *eradication*.

4. Cities and towns offer new ecologies for malaria. For an account of flows of water and natural changes in cities, see Stephanie Pincetl, "Nature, Urban Development and Sustainability—What New Elements Are Needed for a More Comprehensive Understanding," *Cities* 29 (2012): S32–S37.

5. Personal communication from Ministry of Health workers in Bahir Dar and neighborhood residents in Kebele, January 10, 2012.

6. Gregory S. Noland et al., "Surveillance as an Intervention for Malaria: Response to Potential Outbreaks Identified through District-Level Surveillance in Amhara Region, Ethiopia" (paper presented at ASTMH Conference, Atlanta, November 2012). My thanks to Joseph Malone and Gregory Noland for alerting me to this 2012 malaria outbreak and the Carter Center role in addressing the impact.

7. See Richard Feachem and Oliver Sabot, "A New Global Malaria Eradication Strategy," *Lancet* 371 (2008): 1633–35. Also see Simon I. Hay et al., "The Global Distribution and Population at Risk of Malaria: Past, Present, and Future," *Lancet Infectious Diseases* 4, no. 6 (2004): 327.

8. James L. A. Webb Jr., *Humanity's Burden: A Global History of Malaria* (New York: Cambridge University Press, 2009), 166–70. Webb cites Africa's exclusion from the WHO eradication plan as that continent being "sidelined" in its 1955 World Health Assembly.

9. "Malarial Control Conference Suggests Eradication Plan," *Ethiopian Herald*, March 1, 1958.

10. On recurrent epidemics over time, see K. Negash et al., "Malaria Epidemics in the Highlands of Ethiopia," *East African Medical Journal* 82, no. 4 (April 2005): 186–92; and Asnakew Kebede, "Overview of the History of Malaria Epidemics in Ethiopia" (paper presented at the Workshop on Capacity Building on Malaria Control in Ethiopia, Addis Ababa, May 4–9, 2002).

11. A good summary of the malaria/smallpox comparison is "Eradication Efforts: Malaria vs. Smallpox," http://www.uniteforsight.org/global-health-history/module 4.

12. S. L. Takala and C. V. Plowe, "Genetic Diversity and Malaria Vaccine Design, Testing and Efficacy: Preventing and Overcoming Vaccine Resistant Malaria," *Parasite Immunology* 31, no. 9 (2009): 560, 568.

13. Gretchen Vogel, "More Sobering Results for Malaria Vaccine," *ScienceNow* (March 20, 2013), http://news.sciencemag.org/2013/03 more-sobering-results-malaria-vaccine.

14. Personal communication at WHO Headquarters, Geneva (July 2011), from Ambachew Medhin, who was the prospective director of the Ethiopia trial.

15. Vogel, "More Sobering Results." Dr. Gregory Noland, malaria specialist of the Carter Center, has pointed out to me that the vaccine for those few who get some immunity will last only a year.

16. Feachem and Sabot, "New Global Malaria Eradication Strategy," 1633.

Bibliography

Archives
British Library MS, (OR 534) f. 156r
Hiob Ludolf Centre (University of Hamburg)
n.a. *Metshaf Senksar Ba Ge'ez na Amaregna.* n.d.
Dawit Psalter, mss. Smo11. Digital Oriental Manuscript Library.

World Health Organization (Geneva)
Parasitology Collection

WHO7 Ethiopia 1937-1942 JKT I WHO7.0011
ETH 1937
Office International d'Hygiene Publique. "Epidemiologie du Paludisme dans la Region du Lac Tana." Note presentee par S.E. le Professeur Castellani, Delegue de Colonies Italiennes. No. 177, 18 October 1938. Session d'Octobre 1938.
Office International d'Hygiene Publique. "Le Paludisme dans la Region Uollo-Jeggu Etiopie_ pendant la Saison des Pluies." Note presentee par S.E. le Professeur Castellani, Delegue de Colonies Italiennes. No. 176, 18 October 1938. Session d'Octobre 1938.

Ethiopia 1937–1942 JKT I WHO7.0012
ETH 1943–1946
Melville, A. R., Duncan B Wilson, J. P. Glasgow, and K. S. Hocking. "Malaria in Abyssinia." *East African Medical Journal* 22 (1945): 285–94.
ETH 1948–1950
Giaquinto-Mira, Mario. "Notes on the geographical distribution and biology of ANOPHELINAE and CULICINAE in Ethiopia." Imperial Ethiopian Medical Research Institute, Addis Ababa, October 1948.
ETH 1954–1956
Covell, Gordon. "Report on Health Conditions at the Proposed Site for the Construction of a City in the Southern End of Lake Tana, Ethiopia." 8–26 October 1952. 30.8.55 Malaria section.
Covell, Gordon. Report on a Visit to Ethiopia September–October 1955." January 1956 (typescript copy in 1943–1956 WHO7.0012).
Farid, M. A. "Report on a Visit to Ethiopia 12–20 August 1956." WHO typescript.
ETH 1957–1958
WHO Assistance to Ethiopia in the Field of Malaria (9.11.1958).

ETH 1958–1959
WHO Pilot Malaria Project 1956–1959 Sketch Map of Project Zone n.d. Najjar, Abdallah. Dembia Pilot Project, Ethiopia. 22 October 1959.
ETH 1964–1965
Malaria Eradication Service, *YaWoba MaTafiya Dirgit*. Addis Ababa: *Yatemertna Ya Mastawaqiya* Division, June 1964. [Amharic typescript]
ETH 1968–1969
Report on a Visit to Ethiopia 10–18 December 1969.
Dr. S. C. Edwards, WHO regional Public Health Administrator (Malaria) January 1970.

WHO7.0013
ETH 1970–1971
Mal Ethiopia n.d. "Chronological Review" included in file with no author.
"Report of a Strategy Review Team, Ethiopia. May 6–27, 1970."
ETH 1972
"Report of An Independent Malaria Review Team in Ethiopia" 24 May–2 June ML/370/19(3) 1972.
ETH 1972
Information on Entomological Activities of the National Malaria Eradition Service of the Imperial Ethiopian Government. November 1972.
K. F. Schaller, W. Kuls, Athiopie-Ethiopa A Geomedical Monograph.

WHO7.0014
ETH 1991
Awash Teklahaimanot, "Report of a Mission to Ethiopia 4 August–13 September 1991: Malaria Epidemics in Ethiopia with Particular Reference to War-affected Northern Regions."
Awash Teklahaimanot. "Report of a Follow-up Mission to Ethiopia, 27 October–2 November 1991; Malaria Epidemics in Ethiopia with Particular Reference to the War-affected Northern Regions."

WHO7.0774
PARAS1-Afro-Ethiopia (Maps) 1936–1982 Fol 2.
Africa Orientale Italiana. Carta Dimonstativa delle Zone Malariche 1936.
Provisional Malaria Map of East Africa (Abyssinia), 1943.

Bound volumes (WHO Library — Geneva)
World Health Organization Regional Office for the Eastern Mediterranean (EM/MAL/1-EM/MAL/17) 1951–1955.
H. G. S. Morin, Malaria Problems in Eastern Mediterranean Region Preliminary Remarks, March 1951.
Dr. D. Sonti (Malaria consultant, EMRO), "The Present Status of Malaria in Ethiopia (Report on Visit to Ethiopia) 10–21 May 1954. EM/ME-Tech.2/1-54."
Second Regional Conference on Malaria Eradication Addis Ababa, 16–21 November 1959.
Dowling, M. A. C. "Review of the Malaria Programme in the African Region."

Presented at the Malaria Eradication Technical Meeting, Brazzaville, 30 November 1959.

The World Health Organization Regional Office for the Eastern Mediterranean (EM/MAL/38-48 1959–1963).

Report on the Second Regional Conference on Malaria Eradication, Addis Ababa, 16–21 November 1959.

"2.1.1 The 1958 Malaria Epidemic in Ethiopia"

Catalog of the Material

WHO Archives (Geneva) on Microform from

L'Office International D'Hygiene Publique

and

The United Nations Relief and Rehabilitation Administration

Files: Microfiche #

Malaria

    —prophylactic measure against . . . 1916   A31

    —report on "Norme per la Profilassi Antimalrica, 1918   A47

    —. . . In Italy and Belgium, 1916   A54

Office International d'Hygience Publique List of Files and

Their Contents

    —in Italian East Africa Eritrea   A4

    —1933–38 Italian East Africa   A4

Health Mission Activities—Ethiopia   H4/4/1

                                                                                          Microfiche #

    —Memorandum from UNRRA Cairo to UNRRA Washington
      advice on the establishment of the Ethiopian Mission   Folder no. 1

    —Initial health programme of the Ethiopian
      mission, March 1946   Folder 1 and 2

    —cable informing of the decision to prolong UNRRA
      Ethiopian health services three months beyond the
      end of 1946; letter dealing with possibility for WHO
      to take over UNRRA health activities in Ethiopia;
      report on a meeting with Dr. Pridie on 18 Sept. 1946   Folder No. 3

Malaria

First Generation of files, 1946–1950 (WHO.1)

and

Second generation of files, 1950–1955 (WHO.2)

Ethiopia, 2-2-28, 4-1-4, 468-1-1, 469-8-19, DC 15-2, DC REL FEV 8, DC TB 7, EQ 9-1, EQ 9-2/24, EQ 10,, EQ 10/1, L 9-2/16, PH 2-5/20 Constitution, ratification 3-2-29.

Malaria,

    —Advisory Board, 453-4-4

    —agriculture, and 453-4-21, 453-5-6, AS6-5

    —anti-malarial drugs, 453-5-1, 453-5-2, 453-5-6, 453-5-7, 453-5-14

    —anti-malarial organization, 453-9-3

    —International Congress, Washington 1948   453-1-4

    —campaigns, 453-1-8, 453-8-1, 453-12-1, AS 8/1

    —conference, DC MAL 20/1, to DC MAL 20/3, OD 20-3/1

    —control programme AS 6-5, CC 4-3/2

    —insecticides, 453-4-6, 453-5-6, 453-5-7, 453-5-8, 453-5-14

Third Generation of files, 1955–1983 (WHO.3)

M2		Malaria
1965–1972	Meetings of the Inter-Country Malaria Eradication Board	M2/86/8
1956	Second Africa Malaria Conference (Annexes)	M2/87/11 J.1*,2.
1958–1970	Economic and Social Losses due to Malaria	M2/180/4
1960–1971	Registration of Areas where Malaria has been eradicated Programme for Malaria Eradication	M2/180/11
1955–1970	African area	M2/372/3/AF J.1 to 6
1956–1959	African area –Special Appraisal of Residual Spraying 1960	M2/372/3/AF J.1*, 2*, 3*

Reports on Malaria (in Africa) 1955, single year snapshots

1960–1961	Investigation of the Importance of Population Movements in Malaria Eradication	M2.445/10*
1962–1962	Studies on Criteria for Malaria Eradication	M2/445/20*

Ethiopia

A18/181/14/A, E11/87/3/ETH, E17/180/2/ETH, H2/181/7, L2/371/2/ETH

Ethiopian Nutritional Institute, Addis Ababa	M3/181/13
Eradication Programmes, Malaria, 1967	M2/87/30

Correspondence   ETH-MPD-002 Regional Director, EMRO to Mr. B. Feinstein, WHO Entomologist, Addis Ababa, 1 November 1976.
Report on field tour 19–20 October 1976.

———

A. H. Taba, M. D. Regional Director to Dr. S. Street, WHO Representative, Addis Ababa 28 January 1976.
A. H. Taba, M.D. to Dr. K Lassen, WHO Malariologist, Addis Ababa.
A. H. Taba to Dr. K. Lassen, WHO malariologist, Addis Ababa. 11 February 1975.
A. H. Taba to Dr. K Lassen, WHO malariologist, Addis Ababa, 30 January 1974.
A. H. Taba, Regional Director to Dr. A. Mohtadi, Senior WHO Malariologist, Addis Ababa, 9 July 1970.

M2/372/3 7 May 1964.
Project Malaria Pre-eradication Programme, 17 March 1964.

M2/372/3
Dr. G. Sambasivan, Director ME to Regional Malaria Advisor, 10 September 1965.

M2/372/3 (A) ETH

J. S. Prince M.D. to Ato Yohannes Tseghe, Minister of State for Health, Imperial
    Ethiopian Government, 25 August 1966.
Report of Malaria Review Committee of 12 August ; 25 February 1965.
Plan of Operation for a Malaria Pre-Eradication Programme in Ethiopia on
    18 January 1965.

M2/372/3(a) ETH
From E. B. Weeks, Chief Epidemiological Assessment, Division of Malaria
    Eradication to Chief, Planning and Programme, Malaria Eradication,
    17 January 1964
Subject: Plan of Action on the Malaria Eradication Programme in Ethiopia
Regional Director, EMRO, Alexandria To Director of ME, WHO Headquarters,
    8 January 1964

M2/372/3 (a)
C. A. Alvarado,M.D. Regional Director Malaria Eradication to: Dr. E. B. Weeks,
    EMRO
"Revised Plan of Operation for A Malaria Eradication Programme of the Imperial
    Ethiopian Government" (USAID and WHO) 9 May 1967, from Minister of
    Public Health Bitwoded Asfaha W. Michael, 16 August 1963.
Map "Strata for May 1969 SMPS."
Samuel Putnam, M.D. to Dr. G. Gramiccia, Chief, Epidemiological Assessment/
    Division of Malaria Eradication, WHO, 27 March 1969.
E. Shafa, M.D. M.P. H. "Epidemiological Investigation of a Malaria Epidemic in
    Begemedir and Semien Province of Ethiopia."

M2/372/3 ETH
Information provided by EMRO to Dr.Alvarado during his visit to EMRO,
    November–December 1961

M2/372/11 Ethio
Unsigned confidential memo from EMRO dated 21 August 1959

M2/372/3
February 21, 1958 (By I. Najjar, parasitologist)
Report and Suggestions of Meetings Reviewing Present Malaria Situation in Ethiopia

M2/372/ (MPD001)
Quarterly report for Jan–March 1965.
Opening of Second Session of Malaria Eradication Training on 23 January 1960.

M2/372/3
Quarterly Report, 1 July to 30 September 1964.
Reports arrival of 108 USAID vehicles consisting of 65 half-ton Jeep C-J 3B, 42 Pick-
    Jeep Wagoneers, and 1 Jeep Wagoneer.

M2/372/3 (b) ETH
8 April 1964 Quarterly Report 1 January to 31 March 1964.

M2/372/3b) ETH
Annual Report of the Residual Spraying Campaign-1960.

M2/372/3 (b) ETH
7 September 1959 Brief Report of the Malaria Pilot Project in 1958–1959.

M2/372/3/(b) ETH
31 December 1965.

M2/372/3 (b) ETH
Annual Report for the FY 195M2/374/3(b), April 25 1966.

M2/374/3 (b)
Assignment Report Malaria Eradication Programme in Ethiopia July 1968–July 1969.

M2/372/2(B) ETH
Assignment Report Malaria Eradication Programme in Ethiopia July 1967–July 1968 by Dr. S. C. Luen, WHO Epidemiologist.
Report on a Visit to Ethiopia 10–18 December 1969 Dr. S. C. Edwards, WHO Regional Public Health Administrator (Malaria).

Articles and Books

Abebe Getahun and Eshete Dejen. *Fishes of Lake Tana: A Guide Book*. Addis Ababa: Addis Ababa University Press, 2012.
Abeku, T. A., S. J. De Vlas, G. J. J. Borsboom, A. Tadege, Y. Gebreyesus, H. Gebreyohannes, D. Alamirew, A. Seifu, N. J. D. Nagelkerke, and J. D. F. Habbema. "Effects of Meteorological Factors on Epidemic Malaria in Ethiopia: A Statistical Modeling Approach Based on Theoretical Reasoning." *Parasitology* 128, no. 6 (2004): 585–93.
Ameneshewa, B, and M. W. Service. "The Relationship between Female Body Size and Survival Rate of the Malaria Vector *Anopheles arabiensis* in Ethiopia." *Medical Veterinary Entomology* 10, no. 2 (1996): 170–72.
Amorosa, Louis F., Gilberto Corbellini, and Mario Coluzzi."Lessons Learned from Malaria: Italy's Past and Sub-Sahara's Future." *Health and Place* 11, no. 1 (2005): 67–73.
Asnakew Kebede. "Overview of the History of Malaria Epidemics in Ethiopia." Paper presented at the Workshop on Capacity Building on Malaria Control in Ethiopia. Addis Ababa, May 4–9, 2002.
———. "Spatial Analysis of Malaria Incidence at Village Level in Areas with Unstable Transmission in Ethiopia." MSc. thesis, Department of Geography, Boston University, 2008.
Asnakew Kebede, James C. McCann, Anthony E. Kiszewski, and Yemane Ye-Ebiyo. "New Evidence of the Effects of Agro-Ecologic Change on Malaria Transmission." *American Journal of Tropical Medicine and Hygiene* 73, no. 4 (2005): 676–80.
Assefa Tulu, N. "Malaria." In *The Ecology of Health and Disease in Ethiopia*, edited by Zein A. Zein and Helmut Kloos, 341–52. Boulder, CO: Westview Press, 1993.
Aylward, Bruce, K. A. Hennessey, N. Zagaria, J. M. Olivé, and S. Cochi. "When Is a Disease Eradicable? 100 Years of Lessons Learned." *American Journal of Public Health* 90, no. 10 (2000): 1515–20.
Baker, Samuel W. *The Nile Tributaries of Abyssinia and the Sword Hunters of the Hamran Arabs*. London: Macmillan, 1874.
Balkew, Meshesha, Ibrahim Muntaser, Lizette L. Koekemoer, Basil D. Brooke, Howard Engers, Abraham Aseffa, Teshome Gebre-Michael, and Ibrahim Elhassen.

"Insecticide Resistance in *Anopheles arabiensis* (Diptera: Culicidae) from Villages in Central, Northern, and South West Ethiopia and Detection of *kdr* Mutation." *Parasites and Vectors* (2010): 3:40.

BBC News. "'Double Whammy' Malaria Drug Hope." April 11, 2009. http//news.bbc. co.uk/gp/pr/fr/-/2/hi/health.

Bean, William B. "The Illness of Charles Darwin." *American Journal of Medicine* 65, no. 4 (1978): 572–74.

Bezawit Eshetu. "Nutrient Composition of Maize Pollen and Its Microbial Degradation." MSc thesis, Biology Department, Addis Ababa University, 2007.

Bosworth, C. E., E. van Donzel, B. Lewis, and Ch. Pellat, eds. *The Encyclopaedia of Islam.* Vol. 6. Leiden: Brill, 1991.

Bourke, Dermot R. W. *Sport in Abyssinia; or, The Mareb and Tackazzee.* London: Murray, 1876.

Brambilla, A. "Il problema della malaria a Dire Daua," *Rivista di Malariologia* 19, no. 5 (1940): 290–309.

———. "L'anofelismo nella zona di Dire Daua (Harar): Prima nota." *Rivista dalla Malariologia* 20 (1941): 1–25.

Bruce, James. *Travels to Discover the Source of the Nile, in the Years 1768, 1769, 1770, 1771, 1772, and 1773.* 5 vols. Edinburgh: Ruthven, 1790.

Bruce-Chwatt, Leonard J. "Lessons Learned from Applied Field Research Activities in Africa during the Malaria Eradication Era." Supplement, *Bulletin of the World Health Organization* 62 (1984): 19–29.

———. "Malaria Eradication at the Crossroads." *Bulletin of the New York Academy of Medicine* 45, no. 10 (October 1969): 999–1012.

Burton, Richard F. *First Footsteps in East Africa; or, An Exploration of Harar.* Edited by Gordon Waterfield. New York: Praeger, 1966.

Carlson, Andrew J., and Dennis G. Carlson. *Health, Wealth, and Family in Rural Ethiopia: Kossoye, North Gondar Region, 1963–2007.* Addis Ababa: Addis Ababa University Press, 2008.

Castellani, Aldo. "Épidémiologie du Paludisme dans le Région du Lac Tsana: Note présentée par S.E. le Professeur Castellani, Délégé des Colonies Italiennes." Ethiopia 1937–1942 JKT 1 WHO 7.0011.

Centro Studi per l'Africa Orientale Africana. *Missione di Studio Lago Tana.* 7 vols, Rome: 1938.

Charlwood, J. D., J. Kihonda, S. Sama, P. F. Billingsley, H. Hadji, J. P. Verhave, E. Lyimo, P. C. Luttikhuizen, and T. Smith. "The Rise and Fall of *Anopheles arabiensis* (Diptera: Culicidae) in a Tanzanian Village." *Bulletin of Entomological Research* 85, no. 1 (1995): 37–44.

Consociazione Turistica Italiana. *Africa Orientale Italiana.* Milan: Consociazione Turistica Italiana, 1938.

Conway, Declan, and Alan Dixon. *The Hydrology of Wetlands in Illubabor Zone in Sustainable Wetland Management in Illubabor Zone, South-west Ethiopia.* Report 1 for Objective 2 (Report 2 of 9). (July 2000).

Coogle, C. P. "The Spleen Rate as a Measure of Malaria Prevalence in the United States." *Public Health Reports* 42, no. 25 (1927): 1683–88.

Corradetti, Augusto. "La biologia dell'Anopheles gambiae e il problema malarico ell'Africa Orientale Italiana." *Rivista di Biologia Coloniale* 2, no. 5 (1939): 321–27.

————. "La malaria nella regione Uollo-Jeggiu nel periodo Luglio-Ottobre 1937." *Bolletino della Societa' Italiana di Biologica Sperimentale* 13, no. 2 (1938): 115–16.

————. "L'anofelismo nella regione del Semien durante la stagione secca." *Bolletino della Societa' Italiana di Biologica Sperimentalel* 13, no. 2 (1938): 114–15.

————. "L'anofelismo nella regione Uollo Jeggiu." *Rivista di Parassitologia* 3, no. 3 (1939): 207–19.

————. "Le conoscenze sulla distribuzione delle species Anofeliche nell'Africa Orientale Italiana." *Rivista di Biologia Coloniale* 3, no. 4 (1940): 419–29.

————. "Notizie preliminari sulla fauna Anofelica della regione di Gondar del Lago Tana e della regione del Simien." *Rivista di Parassitologia* 3, no. 2 (1939): 153–56.

————. "Ricerche epidemiologiche sulla malaria nella regione Uollo-Jeggiu durante la stagione delle piogge." *Rivista di Malariologia* 17 (1938): 101–10.

————. "Ricerche sulla biologia dell Anopheles (Mysomyia) gambiae." *Rivista di Parassitologia* 2, no. 2 (1938): 143–50.

————. *Rivista di Malariologia* 19, no. 1 (1940): 39 ff.

————. "Sui fattori determinanti i tipi epidemici della terzana benigna de della estivo-autunnale." *Rivista di Malariologia* 18, no. 1 (1939): 3–10.

————. "Sulla fauna anofelica della regione amarica." *Bolletino della Societa' Italiana di Biologica Sperimentale* 14, nos. 6–7 (1939): 352.

————. *Una specie asiatica di "Anopheles" rinvenuta in Etiopia ("A. dthali Patton," 1905).* Rome: Accademia Nazionale dei Lincei, 1937.

Covell, Gordon. "Malaria in Ethiopia." *Journal of Tropical Medicine and Hygiene* 60 (January 1957): 7–16.

Cox, Francis E. G. "History of the Discovery of the Malaria Parasites and Their Vectors." *Parasites and Vectors* 3 (2010): 5–13.

Cutler, Sally J. "Possibilities for Relapsing Fever Reemergence." *Emerging Infectious Diseases* 12, no. 3 (March 2006): 369–74.

Davidson, George. "*Anopheles gambiae*: A Complex of Species." *Bulletin of the World Health Organization* 31, no. 5 (1964): 625–34.

————. "*Anopheles gambiae* Complex." *Nature* 196 (1962): 907.

De Cosson, Emilius A. *The Cradle of the Blue Nile: A Visit to the Court of King John of Ethiopia.* London: Murray, 1877.

Delenasaw Yewhalaw, Wim Van Bortel, Leen Denis, Marc Coosemans, Luc Duchateau, and Niko Speybroeck. "First Evidence of High Knockdown Resistance Frequency in *Anopheles arabiensis* (Diptera: Culicidae) from Ethiopia." *American Journal of Tropical Medicine and Hygiene* 83, no. 1 (2010): 122–25.

Delenasaw Yewhalaw, Fantahun Wassie, Walter Steurbaut, Pieter Spanoghe, Wim Van Bortel, Leen Denis, Dejene A. Tessema, et al. "Multiple Insecticide Resistance: An Impediment to Insecticide-Based Malaria Vector Control Program." *PLoA ONE* 6, no. 1 (2011).

Desta Teklewold. *Yamarigna Mazigaba Qallat.* Addis Ababa: Artistic Printers, 1970.

Diatta, M., A. Spiegel, L. Lochouarn, and D. Fontenille. "Similar Feeding Preferences of *Anopheles gambiae* and A. *arabiensis* in Senegal." *Transactions of the Royal Society of Tropical Medicine and Hygiene* 92 no. 3 (1998): 270–72.

Dingle, Ceri. "The Great Malarial Bed-Net Swindle," April 28, 2009. http://www.spikedonline.com/index.php.

Donnelly, Martin J., Monica C. Licht, and Tovi Lehmann. "Evidence for Recent Population Expansion in the Evolutionary History of the Malaria Vectors *Anopheles arabiensis and Anopheles gambiae.*" *Molecular Biology and Evolution* 18, no. 7 (2001): 1353–64.

Evans, Andrew G., and Thomas E. Wellems. "Coevolutionary Genetics of *Plasmodium* Malaria Parasites and Their Human Hosts." *Integrative and Comparative Biology* 42, no. 2 (April 2002): 401–7.

Feachem, Richard, and Oliver Sabot. "A New Global Malaria Eradication Strategy." *Lancet* 371 (2008): 1633–35.

Fontaine, Russell E., Abdallah E. Najjar, and Julius S. Prince. "The 1958 Malaria Epidemic in Ethiopia." *American Journal of Tropical Medicine and Hygiene* 10, no. 6 (1961): 795–803.

Fuller, Thomas. "Spread of Malaria Feared as Drug Loses Potency." *New York Times,* January 26, 2009. http://www.nytimes.com/2009/01/27/health/27malaria.html.

Gabaldon, Arnoldo. "Global Eradication of Malaria: Changes of Strategy and Future Outlook." *American Journal of Tropical Medicine and Hygiene* 18, no. 5 (1969): 641–56.

Gallagher, James. "GM Mosquitoes Offer Malaria Hope." *BBC News,* April 20, 2011.

Galvin, Kathleen T., Nick Petford, Frances Ajose, and Dai Davies. "An Exploratory Qualitative Study on Perceptions about Mosquito Bed Nets in the Niger Delta: What Are the Barriers to Sustained Use?" *Journal of Multidisciplinary Healthcare* 4 (2011): 73–83.

Gamst, Frederick. "A Note on a Malevolent Malaria Spirit and Its Significance to Public Health Workers." *Journal of Health* 6, no. 1 (1966): 24–25.

Gebre-Mariam, Negussie, Yahya Abdulahi, and Assefa Mebrate. "Malaria." In *The Ecology of Health and Disease in Ethiopia,* edited by Zein A. Zein and Helmut Kloos, 136–50. 1st ed. Addis Ababa: Ministry of Health, 1988.

Gebremariam, Messay Fettene, Patrick Bitsindou, Magaran Bagayoko, and Lucien Manga. *Entomological Profile of Malaria in Ethiopia.* Federal Ministry of Health, Ethiopia. World Health Organization, September 2007.

Giaquinto Mira, Mario. "La lotta antimalarica in A.O. I." In *Opere per l'organizzazione civile in Africa orientale italiana.* Addis Ababa: Servizio Tipografico Governo Generale A.O.I., 1939.

Gish, Oscar. "Malaria Eradication and the Selective Approach to Health Care: Some Lessons from Ethiopia." *International Journal of Health Services* 22, no. 1 (1992): 179–92.

Githeko, Andrew K., Nicholas I. Adungo, Diana M. Karanja, William A. Hawley, John M. Vulule, Isack K. Seroney, Ayub V. O. Ofulla, et al. "Some Observations on the Biting Behavior of *Anopheles gambiae s.s., Anopheles arabiensis,* and *Anopheles funestus* and Their Implications for Malaria Control." *Experimental Parasitology* 82, no. 3 (1996): 306–15.

Govere, J., D. N. Durrheim, L. Baker, R. Hunt, and M. Coetzee. "Efficacy of Three Insect Repellants against the Malaria Vector *Anopheles arabiensis.*" *Medical and Veterinary Entomology* 14, no. 4 (2000): 441–44.

Graham, Douglas C. *Glimpses of Abyssinia; or, Extracts from Letters Written while on a Mission from the Government of India to the King of Abyssinia in the Years 1841, 1842, and 1843.* London: Longmans, Green, 1867.

Grassi, Giovanni B. *Studi di uno zoologo sulla malaria*. Rome: Accademia dei Lincei, 1900.

Graves, Patricia M., Frank O. Richards, Jeremiah Ngondi, Paul M. Emerson, Estifanos Biru Shargie, Tekola Endeshaw, Pietro Ceccato, et al. "Individual, Household and Environmental Risk Factors for Malaria Infection in Amhara, Oromia and SNNP Regions of Ethiopia." *Transactions of the Royal Society of Tropical Medicine and Hygiene* 103, no. 12 (2009): 1211–20.

Guidi, Ignazio. *Vocabolario Amarico-Italiano*. Rome: Istituto per l'Oriente, 1935.

Hackett, Lewis W. *Malaria in Europe: An Ecological Study*. London: Oxford University Press, 1937.

Harris, William C. *The Highlands of Ethiopia*. 3 vols. London: Longman, 1844.

Hay, Simon I., Carlos A. Guerra, Peter W. Gething, Anand P. Patil, Andrew J. Tatem, Abdisalan M. Noor, Caroline W. Kabaria, et al. "A World Malaria Map: *Plasmodium falciparum* Endemicity in 2007." *PLoS Med* 6, no. 3 (March 2009): 286–302.

Hay, Simon I., Carlos A. Guerra, Andrew J. Tatem, Abdisalan M. Noor, and Robert W. Snow. "The Global Distribution and Population at Risk of Malaria: Past, Present, and Future." *Lancet Infectious Diseases* 4, no. 6 (2004): 327–36.

Henderson, Donald A. "Eradication: Lessons from the Past." Supplement 2, *Bulletin of the World Health Organization* 76 (1998): 17–21.

Hölldobler, Bert, and Edward O. Wilson. *The Ants*. Cambridge, MA: Harvard University Press, 1990.

———. *The Leafcutter Ants: Civilization by Instinct*. New York: Norton, 2011.

———. *The Superorganism: The Beauty, Elegance, and Strangeness of Insect Societies*. New York: Norton, 2009.

Holling, C. S. "Understanding the Complexity of Economic, Ecological, and Social Systems." *Ecosystems* 4 (2001): 390–405.

Humphries, Courtney. "An Evolving Foe: Applying Genomic Tools to the Fight against Malaria." *Harvard Magazine* 112, no. 4 (March–April 2010): 42–46.

Istituto Agricolo Coloniale. *Main Features of Italy's Action in Ethiopia, 1936–1941*. Florence: Istituto Agricolo Coloniale, 1946.

Johnston, Charles. *Travels in Southern Abyssinia, through the Country of Adal to the Kingdom of Shoa during the Years 1842–43*. London: Madden, 1844.

Jones, Kate E., Nikkita G. Patel, Marc A. Levy, Adam Storeygard, Deborah Balk, John L. Gittleman, and Peter Daszak. "Global Trends in Emerging Infectious Diseases." *Nature* 451 (February 2008): 990–93.

Kane, Thomas L. *Amharic-English Dictionary*. Wiesbaden: Harrassowitz, 1990.

Kassahun Kebede, H. "The Social Dimension of Development-Induced Resettlement: The Case of Gilgel Gibe Hydroelectric Dam in Southwest Ethiopia." In *People, Space and the State: Migration, Resettlement and Displacement in Ethiopia*, edited by Alula Pankhurst and Francois Piguet. Addis Ababa: Ethiopian Society of Sociologists, Social Workers and Anthropologists, 2004.

Keeley, James, and Ian Scoones. "Knowledge, Power and Politics: The Environmental Policy-Making Process in Ethiopia." *Journal of Modern African Studies* 38, no. 1 (2000): 89–120.

Kloos, Helmut. "Primary Health Care in Ethiopia under Three Political Systems: Community Participation in a War-Torn Society." *Social Science and Medicine* 46, nos. 4–5 (1998): 505–22.

Koch, Magaly, and James C. McCann. "Satellite Imagery, Landscape History, and Disease: Mapping and Visualizing the Agroecology of Malaria in Ethiopia." Working Paper 8, PSAE Research Series. Boston University, 2010.

Lamb, Henry F., C. Richard Bates, Paul V. Coombes, Michael H. Marshall, Mohammed Umer, Sarah J. Davies, and Eshete Dejen. "Late Pleistocene Desiccation of Lake Tana, Source of the Blue Nile." *Quaternary Science Reviews* 26, nos. 3–4 (2007): 287–99.

Larebo, Haile M. *The Building of an Empire: Italian Land Policy and Practice in Ethiopia, 1935–1941.* New York: Oxford University Press, 1994.

Lefèvre, Thierry, Louis-Clément Gouagna, Kounbrobr Roch Dabiré, Eric Elguero, Didier Fontenille, François Renaud, Carlo Costantini, and Frédéric Thomas. "Beyond Nature and Nurture: Phenotypic Plasticity in Blood-Feeding Behavior of *Anopheles gambiae* s.s. when Humans Are Not Readily Accessible." *American Journal of Tropical Medicine and Hygiene* 81, no. 6 (2009): 1023–29.

Leslau, Wolf. *Concise Amharic Dictionary.* Berkeley: University of California Press, 1996.

Levine, Donald N. *Wax and Gold: Tradition and Innovation in Ethiopian Culture.* Chicago: University of Chicago Press, 1972.

Lindsay, S. W., and W. J. Martens. "Malaria in the African Highlands: Past, Present and Future." *Bulletin of the World Health Organization* 76, no. 1 (1998): 33–45.

MacDonald, George. *The Epidemiology and Control of Malaria.* Oxford: Oxford University Press, 1957.

malEra Consultative Group on Monitoring, Evaluation, and Surveillance. "A Research Agenda for Malaria Eradication: Monitoring, Evaluation, and Surveillance." *PLoS Med* 8, no. 1 (2011). www.plosmedicine.org.

Mboera, Leonard E. G., Kesheni P. Senkoro, Benjamin K. Mayala, Susan F. Rumisha, Rwehumbiza T. Rwegoshora, Malongo R. S. Mlozi, and Elizabeth H. Shayo. "Spatio-Temporal Variation in Malaria Transmission Intensity in Five Agro-Ecosystems in Mvomero District, Tanzania." *Geospatial Health* 4, no. 2 (2010): 167–78.

McCann, James C. *From Poverty to Famine in Northeast Ethiopia: A Rural History, 1900–1935.* Philadelphia: University of Pennsylvania Press, 1987.

———. *Maize and Grace: Africa's Encounter with a New World Crop, 1500–2000.* Cambridge: Harvard University Press, 2009.

———. *People of the Plow: An Agricultural History of Ethiopia, 1800–1990.* Madison: University of Wisconsin Press, 1995.

McCullough, David. *John Adams.* New York: Simon and Schuster, 2001.

McNeil, Donald G., Jr. "In Ethiopia's Malaria War, Weapons Are the Issue." *New York Times*, December 9, 2003.

———. "The Soul of a New Vaccine." *New York Times*, December 11, 2007.

Melville, A. R., Duncan B. Wilson, J. P. Glasgow, and K. S. Hocking. "Malaria in Abyssinia." *East African Medical Journal* 22 (September 1945): 285–94.

Mengistu, M., M. Maru, and Z. Ahmed. "Malaria in Gondar, Ethiopia, 1975–1978: A Review of 435 Cases with Special Emphasis on Cerebral Malaria." *Ethiopian Medical Journal* 17, no. 3 (1979): 57–62.

Mercier, Jacques. *Asrès, le magicien éthiopien: Souvenirs 1895–1985.* Paris: Lattès, 1988.

Michener, James A. *Tales of the South Pacific.* New York: Macmillan, 1947.

Minakawa, Noboru, Stephen Munga, Francis Atieli, Emmanuel Mushinzimana, Guofa Zhou, Andrew K. Githeko, and Guiyan Yan. "Spatial Distribution of Anopheline Larval Habitats in Western Kenya Highlands: Effects of Land Cover Types and Topography." *American Journal of Tropical Medicine and Hygiene* 73, no. 1 (2005): 157–65.

Minakawa, Noboru, Elizabeth Omukunda, Guofa Zhou, Andrew Githeko, and Guiyan Yan. "Malaria Vector Productivity in Relation to the Highland Environment in Kenya." *American Journal of Tropical Medicine and Hygiene* 75, no. 3 (2006): 448–53.

Minakawa, Noboru, Gorge Sonye, Gabriel O. Dida, Kyoko Futami, and Satoshi Kaneko. "Recent Reduction in the Water Level of Lake Victoria Has Created More Habitats for *Anopheles funestus.*" *Malaria Journal* 7 (2008): 119–25.

Ministry of Health, Ethiopia. *Guidelines for Malaria Epidemic Prevention and Control in Ethiopia.* Addis Ababa: Ministry of Health, July 1999.

Moorehead, Alan. *The Blue Nile.* London: Hamish Hamilton, 1962.

Munga, Stephen, Laith Yakob, Emmanuel Mushinzimana, Guofa Zhou, Tom Ouna, Noboru Minakawa, Andrew Githeko, and Guiyun Yan. "Land Use and Land Cover Changes and Spatiotemporal Dynamics of Anopheline Larval Habitats during a Four-Year Period in a Highland Community of Africa." *American Journal of Tropical Medicine and Hygiene* 81, no. 6 (2009): 1079–84.

Mutuku, F. M., M. N. Bayoh, A. W. Hightower, J. M. Vulule, J. E. Gimnig, J. M. Mueke, F. A. Amimo, and E. D. Walker. "A Supervised Land Cover Classification of a Western Kenya Lowland Endemic for Human Malaria: Associations of Land Cover with Larval *Anopheles* Habitats." *International Journal of Health Geographics* 8 (2009): 19.

Nájera, J. A., R. L. Kouznetsov, C. Delacollette. *Malaria Epidemics: Detection and Control, Forecasting and Prevention.* Geneva: World Health Organization, 1998.

Nchinda, Thomas C. "Malaria: A Reemerging Disease in Africa." *Emerging Infectious Diseases* 4, no. 3 (1998): 398–403.

Negash, K., A. Kebede, A. Medhin, D. Argaw, O. Babaniyi, J. O. Guintran, and C. Delacollette. "Malaria Epidemics in the Highlands of Ethiopia." *East African Medical Journal* 82, no. 4 (April 2005): 186–92.

Noland, Gregory S., Solomon Kibret, Eshetu Sata, Belay Bezabih, Ayeligne Mulualem Tuafie, Zerihun Tadesse, Patricia M. Graves, Paul M. Emerson, and Frank O. Richards. "Surveillance as an Intervention for Malaria: Response to Potential Outbreaks Identified through District-Level Surveillance in Amhara Region, Ethiopia." Paper presented at the ASTMH Conference, Atlanta, November 2012.

Norris, Douglas E. "Mosquito-Borne Diseases as a Consequence of Land Use Change." *Ecohealth* 1 (2004): 19–24.

Novello, Elisabetta. *La bonifica in Italia: Legislazione, credito e lotta alla malaria dall'Unità al fascismo.* Milan: Angeli, 2003.

Nyarango, Peter M., Tewolde Gebremeskel, Goitom Mebrahtu, Jacob Mufunda, Usman Abdulmumini, Andom Ogbamariam, Andrew Kosia, et al. "A Steep Decline of Malaria Morbidity and Mortality Trends in Eritrea between 2000 and 2004: The Effect of Combination of Control Methods." *Malaria Journal* 5 (2006): 33.

O'Connor, Charles T., Jr. "The Distribution of Anopheline Mosquitoes in Ethiopia." *Mosquito News* 27, no. 1 (March 1967): 42–54.

Pachauri, R. K., and A. Reisinger, eds. *Climate Change: Synthesis Report.* Geneva: IPCC, 2007.

Packard, Randall M. *The Making of a Tropical Disease: A Short History of Malaria.* Baltimore: Johns Hopkins University Press, 2007.

Pankhurst, Richard. *Economic History of Ethiopia, 1800–1935.* Addis Ababa: Haile Sellassie I University Press, 1968.

———. *Some Factors Influencing the Health of Traditional Ethiopia.* Addis Ababa: Haile Sellassie University Press, 1966.

Parisis. *L'Abisinnia.* Milan, 1888.

Parkyns, Mansfield. *Life in Abyssinia: Being Notes Collected during Three Years' Residence and Travels in That Country.* London: Murray, 1853.

Parmakelis, A., M. A. Russello, A. Caccone, C. B. Marcondes, J. Costa, O. P. Forattini, M. A. M. Sallum, R. C. Wilkerson, and J. R. Powell. "Historical Analysis of a Near Disaster: *Anopheles gambiae* in Brazil." *American Journal of Tropical Medicine and Hygiene* 78, no. 1 (2008): 176–78.

Pearce, Fred. "Should We Let Scientists Release Mutant Mosquitoes into the Wild to Try to Wipe Out Malaria?" May 14, 2011. http://www.dailymail.co.uk/sciencetech/article-1387191.

Pincetl, Stephanie. "Nature, Urban Development and Sustainability—What New Elements Are Needed for a More Comprehensive Understanding." *Cities* 29 (2012): S32–S37.

Plowden, Walter C. *Travels in Abyssinia and the Galla Country: With an Account of a Mission to Ras Ali in 1848.* London: Longmans, Green, 1868.

Prothero, R. Mansell. "Population Movements and Problems of Malaria Eradication in Africa." *Bulletin of the World Health Organization* 24 (1961): 405–25.

Raffaele, G., and A. Canalis. "Relazione." *Rivista di Malariologia* 16 (1937): 1–58.

Raffaele, G., and G. Lega. "Osservazioni su di un ceppi etiopico di Plasmodium falciparum." *Rivista di Malariologia* 1, no. 5 (1937): 388–97.

Reddy, Michael R., Hans J. Overgaard, Simon Abaga, Vamsi P. Reddy, Adalgisa Caccone, Anthony E. Kiszewski, and Michel A. Slotman. "Outdoor Host Seeking Behaviour of Anopheles gambiae Mosquitoes Following Initiation of Malaria Vector Control on Bioko Island, Equatorial Guinea." *Malaria Journal* 10 (2011): 184.

Roberts, Leslie, and Martin Enserink. "Did They Really Say . . . Eradication?" *Science* 318 (2007): 1544–45.

Ross, Ronald. *The Prevention of Malaria (with Addendum on the Theory of Happenings).* 2nd ed. London: Murray, 1911.

Sachs, Jeffrey D. "The $10 Solution: Malaria Kills 2 Million African Children a Year. Here's a Simple Plan for Saving Their Lives." *Time,* January 4, 2007.

Salt, Henry. *A Voyage to Abyssinia, and Travels into the Interior of That Country.* London: Rivington, 1814.

Scott, James C. *Seeing Like a State: How Certain Schemes to Improve the Human Condition Have Failed.* New Haven, CT: Yale University Press, 1998.

Seyoum, A., and D. Abate. "Larvicidal efficacy of *Bacillus thuringiensis* var. *israelensis* and *Bacillus sphaericus* on *Anopheles arabiensis* in Ethiopia." *World Journal of Microbiology and Biotechnology* 13 (1997): 21–24.

Shillcutt, Samuel, Chantal Morel, Catherine Goodman, Paul Coleman, David Bell, Christopher J. M. Whitty, and A. Mills. "Cost Effectiveness of Malaria Diagnostic Methods in Sub-Saharan Africa in an Era of Combination Therapy." *Bulletin of the World Health Organization* 86 (February 2008): 101–10.

Snowden, Frank M. *The Conquest of Malaria: Italy, 1900–1962*. New Haven, CT: Yale University Press, 2006.

Spielman, Andrew, and Michael D'Antonio. *Mosquito: A Natural History of Our Most Persistent and Deadly Foe*. New York: Hyperion, 2001.

Spielman, Andrew, Uriel Kitron, and Richard J. Pollack. "Time Limitation and the Role of Research in the Worldwide Attempt to Eradicate Malaria." *Journal of Medical Entomology* 30, no. 1 (1993): 6–19.

Stern, Henry A. *Wanderings among the Falashas in Abyssinia: Together with a Description of the Country and Its Various Inhabitants*. London: Wertheim, Macintosh, and Hunt, 1862.

Takala, S. L., and C. V. Plowe. "Genetic Diversity and Malaria Vaccine Design, Testing and Efficacy: Preventing and Overcoming 'Vaccine Resistant Malaria.'" *Parasite Immunology* 31, no. 9 (2009): 560–73.

Tanner, Marcel, and Don de Savigny. "Malaria Eradication Back on the Table." *Bulletin of the World Health Organization* 86, no. 2 (February 2008): 82–83.

Tarekegn, A. Abeku, Sake J. de Vlas, Gerard Borsboom, Awash Teklehaimanot, Asnakew Kebede, Dereje Olana, Gerrit J. van Oortmarssen, and J. D. F. Habbema. "Forecasting Malaria Incidence from Historical Morbidity Patterns in Epidemic-Prone Areas of Ethiopia: A Simple Seasonal Adjustment Method Performs Best." *Tropical Medicine and International Health* 7, no. 10 (October 2002): 851–57.

Tedros Ghebreyesus, A., Mitiku Haile, Karen H. Witten, Asefaw Getachew, Ambachew M. Yohannes, Mekonnen Yohannes, Hailay D. Teklehaimanot, Steven W. Lindsay, and Peter Bypass. "Incidence of Malaria among Children Living Near Dams in Northern Ethiopia: Community Based Incidence Survey." *British Medical Journal* 319 (September 1999): 663–66.

Teklehaimanot. "Chloroquine-Resistant Plasmodium falciparum Malaria in Ethiopia." *Lancet* 2 (1986): 127–31.

Tesema Habte Mikael. *Kasate-Berhan Tesema: Ya'Amareñña Mazgaba Qalat*. Addis Ababa: Artistic Publishing House, 1951.

Tomes, Nancy. *The Gospel of Germs: Men, Women, and the Microbe in American Life*. Cambridge, MA: Harvard University Press, 1998.

Trudel, Ryan E., and Arne Bomblies. "Larvicidal Effects of Chinaberry (*Melia azederach*) Powder on *Anopheles arabiensis* in Ethiopia." *Parasites and Vectors* (2011): 4:72.

Vogel, Gretchen. "More Sobering Results for Malaria Vaccine." *ScienceNow* (March 20, 2013). http://news.sciencemag.org/2013/03/more-sobering-results-malaria-vaccine.

Wade, Nicholas. "Genetic Decoding May Bring Advances in Worldwide Fight against Malaria." *New York Times*, October 3, 2002.

Wang, L. "Malaria in Ethiopia." In *SATACA Review* 11, no. 2 (1966).

Webb, James L. A., Jr. *Humanity's Burden: A Global History of Malaria*. New York: Cambridge University Press, 2009.

———. *The Long Struggle against Malaria in Tropical Africa.* New York: Cambridge University Press, 2014.

White, G. B. "Anopheles Gambiae Complex and Disease Transmission in Africa." *Transactions of the Royal Society of Tropical Medicine and Hygiene* 68, no. 4 (1974): 278–301.

Wilson, D. Bagster. "Implications of Malarial Endemicity in East Africa." *Transactions of the Royal Society of Tropical Medicine and Hygiene* 32, no. 4 (January 1939): 435–65.

Winchester, Simon. *Krakatoa: The Day the World Exploded, August 27, 1883.* New York: HarperCollins, 2003.

World Health Organization. "Hopes and Fears for Malaria." *Bulletin of the World Health Organization* 86, no. 2 (February 2008): 91–92.

———. "Malaria in Africa." *RBM: Roll Back Malaria.* March 2002.

Ye-Ebiyo, Yemane, Richard J. Pollack, Anthony Kiszewski, and Andrew Spielman. "A Component of Maize Pollen That Stimulates Larval Mosquitoes (Diptera: Culicidae) to Feed and Increases Toxicity of Microbial Larvicides." *Journal of Medical Entomology* 40, no. 6 (November 2003): 860–64.

———. "Enhancement of Development of Larval *Anopheles Arabiensis* by Proximity to Flowering Maize (*Zea Mays*) in Turbid Water and When Crowded." *American Journal of Tropical Medicine and Hygiene* 68, no. 6 (2003): 748–52.

Ye-Ebiyo, Yemane, Richard J. Pollack, and Andrew Spielman. "Enhanced Development in Nature of Larval *Anopheles Arabiensis* Mosquitoes Feeding on Maize Pollen." *American Journal of Tropical Medicine and Hygiene* 63, nos. 1–2 (2000): 90–93.

Yilma Mekuria, and Girma Wolde Tsadik. "Malaria Survey in North and North Eastern Ethiopia." *Journal of Ethiopian Medicine* 8 (1970): 201–20.

Zahar, A. R. "Review of the Ecology of Malaria Vectors in the WHO Eastern Mediterranean Region." *Bulletin of the World Health Organization* 50, no. 5 (1974): 427–40.

Zulueta, J. de. "Malaria and Mediterranean History." *Parassitologia* 15 (1973): 1–15.

Index